T0135101

Fire-Resistant Paper

Emerging Materials and Technologies
Series Editor
Boris I. Kharissov

Fire-Resistant Paper

Materials, Technologies, and Applications

Ying-Jie Zhu

CRC Press
Taylor & Francis Group
Boca Raton London New York

CRC Press is an imprint of the
Taylor & Francis Group, an **informa** business

First edition published 2022
by CRC Press
6000 Broken Sound Parkway NW, Suite 300, Boca Raton, FL 33487-2742

and by CRC Press
2 Park Square, Milton Park, Abingdon, Oxon, OX14 4RN

© 2022 Ying-Jie Zhu
CRC Press is an imprint of Taylor & Francis Group, LLC

Reasonable efforts have been made to publish reliable data and information, but the author and publisher cannot assume responsibility for the validity of all materials or the consequences of their use. The authors and publishers have attempted to trace the copyright holders of all material reproduced in this publication and apologize to copyright holders if permission to publish in this form has not been obtained. If any copyright material has not been acknowledged please write and let us know so we may rectify in any future reprint.

Except as permitted under U.S. Copyright Law, no part of this book may be reprinted, reproduced, transmitted, or utilized in any form by any electronic, mechanical, or other means, now known or hereafter invented, including photocopying, microfilming, and recording, or in any information storage or retrieval system, without written permission from the publishers.

For permission to photocopy or use material electronically from this work, access www.copyright.com or contact the Copyright Clearance Center, Inc. (CCC), 222 Rosewood Drive, Danvers, MA 01923, 978-750-8400. For works that are not available on CCC please contact mpkbookspermissions@tandf.co.uk

Trademark notice: Product or corporate names may be trademarks or registered trademarks and are used only for identification and explanation without intent to infringe.

Library of Congress Cataloging-in-Publication Data

Names: Zhu, Ying-Jie, author.
Title: Fire-resistant paper : materials, technologies, and applications /
Ying-Jie Zhu.
Description: First edition. | Boca Raton, FL : CRC Press, 2021. | Series:
Emerging materials and technologies | Includes bibliographical
references and index. | Summary: "While paper is a valued product, the
paper industry contributes to environmental pollution and consumption of
natural resources, and the organic substances out of which traditional
paper is made render it highly flammable and easy to burn. This book
introduces a new technology to develop environmentally friendly
fire-resistant paper using highly flexible ultralong hydroxyapatite
nanowires and discusses applications and potential for
commercialization. This work is aimed at materials scientists, chemical
engineers, industrial chemists, and other researchers from across the
scientific and engineering disciplines interested in development of this
exciting alternative to traditional paper"-- Provided by publisher.
Identifiers: LCCN 2021009053 (print) | LCCN 2021009054 (ebook) | ISBN
9780367700058 (hbk) | ISBN 9780367700065 (pbk) | ISBN 9781003144205
(ebk)
Subjects: LCSH: Fire-resistant paper.
Classification: LCC TS1127 .Z58 2021 (print) | LCC TS1127 (ebook) | DDC
676/.04--dc23
LC record available at https://lccn.loc.gov/2021009053
LC ebook record available at https://lccn.loc.gov/2021009054

ISBN: 978-0-367-70005-8 (hbk)
ISBN: 978-0-367-70006-5 (pbk)
ISBN: 978-1-003-14420-5 (ebk)

Typeset in Times
by Deanta Global Publishing Services, Chennai, India

Contents

Preface

Paper was one of the greatest inventions in the world, and it greatly promoted the rapid development of human civilization. Even in today's electronic information age, paper is still a multi-purpose product that is indispensable to people's daily work and life. However, there are some problems for traditional paper based on cellulose fibers from plants which need to be solved. For example, papermaking consumes a large amount of precious natural resources, such as wood—about 20% of the world's wood is used for papermaking—and another problem is the environmental pollution caused by the papermaking industry. Cellulose is a polysaccharide composed of glucose, which is usually combined with hemicellulose, pectin, and lignin. These organic substances are highly flammable, so traditional paper is highly flammable and easy to burn. In the long history of mankind, fire devoured countless precious paper relics, documents, and books, and turned them into ashes in an instant. This is undoubtedly an immeasurable and irrevocable loss for humankind.

The aim of this book is to comprehensively introduce a new kind of fire-resistant paper based on ultralong hydroxyapatite nanowires from all aspects, including its invention, synthesis, properties, and applications, and to bring this new material and new knowledge to the readers, and to arouse the attention and interest of researchers and readers from different disciplines for more intensive research on this new kind of fire-resistant paper, especially on the research and development of its commercialization and large-scale applications.

More than 20 different kinds of fire-resistant paper based on ultralong hydroxyapatite nanowires are presented in this book. In addition, the applications of these different kinds of fire-resistant paper are also discussed, including fire-resistant specialty paper for protection and permanent safe preservation of important documents, archives, and books; fire-resistant "Xuan Paper" for calligraphy and painting; high-temperature-resistant label paper; highly smooth and glossy fire-retardant paper; applications for fire resistance and heat insulation; fire-resistant paper tape for electric cables and fiber-optic cables; automatic fire alarm fire-resistant wallpaper; multimode anti-counterfeiting, encryption and decryption for secret information; paper for environmental protection, such as recyclable separation paper for the removal of organic solvents and rapid separation of water and oil, water purification, and wastewater treatment; solar energy–driven desalination of seawater; filter paper for polluted air purification and anti-haze face mask; energy-related applications such as high-temperature-resistant separator for advanced lithium ion battery; biomedical applications such as deformable biomaterials, high-performance biomedical paper, bone defect repair and bone regeneration, rapid test paper, and other applications. This book is divided into seven chapters.

Author Biography

 **Ying-Jie Zhu** received his Ph.D. from the University of Science and Technology of China in 1994. From 1994 to 1997, he worked as Assistant Professor and then Associate Professor in University of Science and Technology of China. From 1997 to 2002, he worked as Visiting Scholar at University of Western Ontario (Canada), Alexander von Humboldt Research Fellow at Fritz-Haber-Institut der Max-Planck-Gesellschaft (Germany), and Postdoctoral Fellow at University of Utah and University of Delaware (USA). In 2002, he was selected by the Chinese Academy of Sciences under the Program for Recruiting Overseas Outstanding Talents, and started to work as Full Professor at Shanghai Institute of Ceramics, Chinese Academy of Sciences. In 2007, he was selected under the Program of Outstanding Leader of Shanghai Subject Chief Scientist. His main research interests include nanostructured biomaterials and a new type of fire-resistant paper based on ultralong hydroxyapatite nanowires. He has published more than 370 peer-reviewed journal papers and three book chapters. From 2014 to 2019, he was among the Elsevier's annual list of Most Cited Chinese Researchers for six consecutive years. In addition, he has 65 granted patents.

1 Introduction

1.1. HISTORY OF PAPER

Paper was first invented in ancient China and was one of the greatest inventions in the world, which ended humanity's history of writing with bamboo and wood chips and greatly promoted the development of culture, knowledge, science, technology, and civilization of mankind. Paper is a thin nonwoven material made from plant fibers as the main raw material together with some additives, such as fillers, sizing agents, retention aids, defoamers, and bleaching agents. It is primarily used for writing, printing, artworks, and packaging. It is commonly white because of bleaching. The first papermaking process was documented in China during the Eastern Han period. During the 8th century, Chinese papermaking spread to the Islamic world. By the 11th century, papermaking was brought to Europe. Although precursors such as papyrus (a thick, paper-like material produced from the pith of the Cyperus papyrus plant) in ancient Egypt and other Mediterranean cultures and amate in pre-Columbian Americas existed, these materials were actually the lamination of natural plants and strictly should not be defined as true paper in the real sense. A sheet of papyrus was made from the stem of the plant. The outer rind was first stripped off, and the sticky fibrous inner pith was cut lengthwise into thin strips. The strips were placed side by side on a hard surface with their edges slightly overlapping, and then another layer of strips was laid on top at a right angle, and the layers were hammered together into a single sheet and then dried.[1] Parchment made from animal skins and used principally for writing also cannot be considered as paper.[2] These materials are not considered as paper because they did not possess the main features of paper. The characteristic features of papermaking are: (1) complete separation of the cellulose fibers from each other; and (2) well dispersion of cellulose fibers in water to form pulp.

The history of paper can be traced back to the ancient times more than 2,000 years ago. The literature commonly recognizes Lun Cai from China's Eastern Han dynasty (25–220) as the inventor of paper in the year A.D. 105[3, 4] Figure 1.1 shows the portrait of Lun Cai. However, there are evidences suggesting that paper was invented a few hundred years before Lun Cai.[5] One mainstream view is that the paper made before Lun Cai was poor in quality and not widespread, and Lun Cai improved the papermaking craft based on the experiences of his predecessors. He made the "Cai Hou Paper" with much improved quality using the tree bark, hemp heads, old rags, and fishing nets, and these materials were easily available and cheap. Because of the improved quality and relatively low cost of the "Cai Hou Paper," it gradually spread to many areas and was used by more and more people afterward. Therefore, the "Cai Hou Paper" is considered as the origin of the modern paper. However, with more paper cultural relics being excavated from earlier times than China's Eastern Han

FIGURE 1.1 A portrait of Lun Cai.

dynasty, for example, some from China's Western Han dynasty (202 B.C.–A.D. 8) or even earlier, there has been a debate on whether Lun Cai or someone else who lived in earlier times should be recognized as the inventor of paper.

As a result of the invention of paper, Lun Cai was selected as the most influential 100 persons in history by Michael H. Hart in his book titled *The 100: A Ranking of the Most Influential Persons in History.*[6] In 2007, *Time* magazine selected Lun Cai as one of the best inventors: "The Chinese had written on expensive silk and heavy bamboo slats, but this eunuch of the imperial court created paper out of bark, fishnet, and bamboo, which he pressed thin."[7]

1.2. ROLE OF PAPER IN THE DEVELOPMENT OF HUMAN SOCIETY

The invention of paper brought about the revolution of writing materials for the world. Once paper was widely available, knowledge, culture, science, technology, and other information were quickly and widely disseminated, which promoted the rapid development of human civilization. Even in today's electronic information age with rapid development of science and technology, paper is still a multi-purpose product that is indispensable to people's daily work and life. In the second half of the 20th century, the global paper consumption increased approximately six times.

During the Shang (1600–1050 B.C.) and Zhou (1050–256 B.C.) dynasties of ancient China, documents were ordinarily written on bones or bamboo strips, which were very heavy, awkward, and hard to transport. The light material of silk was sometimes used as a writing medium, but was very expensive. After its invention and spread, paper played an important role in early Chinese written and reading culture.[8] The paper books were light in weight, small in size, and could be carried by hand instead of being transported by cart, as was the case before the use of paper. During the Song dynasty (960–1279), the government produced the world's first-known paper-printed money, or banknote. Paper money was bestowed as gift to government officials in special paper envelopes.[9] During the Tang dynasty (618–907), China became the world leader in book production. In addition, the gradual spread of woodblock printing from the late Tang and Song dynasties further boosted China's lead ahead of the rest of the world.[10] From the 4th century to about 1500, the biggest library collections in China were 3–4 times larger than the largest collections in Europe. The imperial government book collections in the Tang dynasty numbered about 5,000–6,000 titles (89,000 juan) in 721.[10] Paper became central to the three types of art in China—poetry, painting, and calligraphy.

Another way that paper has transformed the human society is by supplementing human memory. In many centuries, paper has been competing with the human mind—and now with the computer—as a preferred means of storing knowledge. Memory systems based on writing have proven advantageous, especially when it is important to retain the information reliably for a long period of time.[11]

1.3. TECHNICAL ASPECTS OF PAPERMAKING

The traditional papermaking process is complex and involves many procedures, including soaking, cooking, pounding, degumming, fiber cutting, pulping, papermaking, drying, etc. Figure 1.2 shows a schematic illustration of the traditional handmade papermaking procedures. About 2,000 years later, modern paper factories are able to complete the whole papermaking process automatically and quickly. However, no matter how improved and optimized modern papermaking technology has become, using cellulose fibers from plants such as trees as raw materials for papermaking has not changed.

1.4. PROBLEMS FACING TRADITIONAL FLAMMABLE PAPER

What happens when paper comes in contact with fire? It only takes a few seconds for paper to quickly burn to ashes. The main component of fibers from trees is cellulose, which is a polysaccharide composed of glucose and is usually combined with hemicellulose, pectin, and lignin. These organic substances are highly flammable, so traditional paper is highly flammable and easy to burn. As the saying goes, "Fire cannot be wrapped up in paper." Fire is the natural enemy of paper. Encountering fire will be a disaster for paper cultural relics, books, and documents. In the long history of mankind, countless precious paper cultural relics and books have been burned in fire. This is also one of the main reasons for the destruction and disappearance of many paper cultural relics, documents, and books in many centuries. This is undoubtedly an immeasurable and irrevocable loss for humankind.

For example, on January 30, 2015, a major fire broke out at the Institute for Scientific Information on Social Sciences Library in Moscow, one of Russia's largest public libraries. Founded in 1918, the library is the biggest social science research

FIGURE 1.2 Schematic illustration of the traditional handmade papermaking procedures.

center in Russia, with about 14 million documents in ancient and modern Eastern European languages, including works dating to the 16th century. Unfortunately, the fire lasted for more than ten hours, and millions of precious ancient books, historic documents, and other materials were destroyed in the fire.

Furthermore, paper produced by the mechanical pulping process contains a significant amount of lignin, a component of wood, which can form a yellow material in the presence of light and oxygen, which is why newspapers and many other types of paper yellow with age. Traditional paper is also at risk of acid decay because cellulose itself produces acidic materials. In addition, papermaking involves the felling of a large number of precious trees and other natural resources and consumes a huge amount of electricity, having a significantly negative impact on the environment. In addition, papermaking also causes environmental pollution. To solve these problems of traditional paper, it is necessary to explore new materials as building material for paper. As a result of considerable effort in recent years, the author's research group developed a novel synthetic method and successfully synthesized highly flexible ultralong hydroxyapatite nanowires.[12] This new nanostructured material provides a possibility for the development of a new kind of fire-resistant paper to tackle the problems of traditional paper.

2 Highly Flexible Ultralong Hydroxyapatite Nanowires

2.1. INTRODUCTION TO HYDROXYAPATITE

Hydroxyapatite ($Ca_{10}(OH)_2(PO_4)_6$), also called "basic calcium phosphate" or hydroxyl calcium phosphate, is a member of the calcium phosphate family, and is of great significance especially for vertebrates, since it is the main inorganic constituent of bone and tooth. The content of hydroxyapatite in human tooth enamel is more than 90 wt.%, and is about 70 wt.% in bone. Because it is a kind of biomaterial in the human body, hydroxyapatite has high biocompatibility, excellent environmental friendliness, and wide biomedical applications.[13-15] Hydroxyapatite has high whiteness, high melting point (~1,650ºC), high-temperature resistance, and nonflammability. Unfortunately, hydroxyapatite is usually as hard and brittle as tooth and bone, and is not suitable for making the soft fire-resistant paper.

Hydroxyapatite is formed in vertebrates through the biomineralization process, which is regulated by a complex physiological system. It is known that the biomineralized hard tissues have well-defined micro- and nano-structures.[13, 14] For example, human dental enamel consists of highly ordered hydroxyapatite nanorods and forms a complex hierarchical ordered structure. Unlike the hydroxyapatite formed in the mild *in vivo* environment, the synthetic conditions of hydroxyapatite materials are much harsher, such as high temperatures (sometimes high pressures), various pH values, organic solvents, etc. This may significantly accelerate the nucleation and crystal growth of hydroxyapatite materials, shorten the preparation time.

Synthetic hydroxyapatite materials have promising applications in various fields such as bone defect repair, drug delivery, and bio-imaging. The rapidly increasing demands for high-performance materials in biomedical and other fields inspire researchers to explore new hydroxyapatite nanostructured materials. Intensive studies have focused on the synthesis, microstructure, morphology, properties, and applications of the as-prepared hydroxyapatite materials. In the last several decades, the studies on hydroxyapatite-based materials have much interest worldwide. Readers may refer to some recent review articles on the synthesis and applications of hydroxyapatite materials.[16-21]

Figure 2.1 shows the rapidly increasing number of papers published each year on hydroxyapatite-based materials from 1995 to 2018, the number of yearly published papers regarding hydroxyapatite-related materials exceeded 4,000 in 2017 and 2019, indicating that hydroxyapatite-based materials have become a hot research topic worldwide.

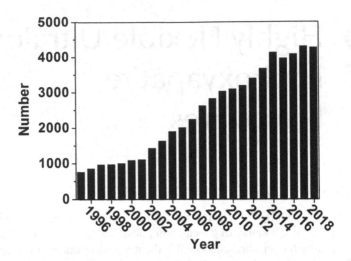

FIGURE 2.1 Number of papers published on hydroxyapatite-based materials each year from 1995 to 2018. The data were obtained from a search in Web of Science using the key word "hydroxyapatite."

2.2. HOW TO SOLVE PROBLEM OF HIGH BRITTLENESS OF HYDROXYAPATITE MATERIALS

Since 2008, the author's research team has done a lot of research work on hydroxyapatite nanostructured materials and their applications. In 2013, the author's research group developed the calcium oleate precursor solvothermal method for the synthesis of ultralong hydroxyapatite nanowires. The ultralong hydroxyapatite nanowires are defined as the hydroxyapatite nanowires with diameters of smaller than 100 nm, lengths of larger than 100 μm, and aspect ratios of higher than 1,000. The lengths of the as-prepared ultralong hydroxyapatite nanowires were several hundred of micrometers, and the diameters of nanowires were only about 10 nm, which is equivalent to about 1/10,000 of the thickness of a human hair. The aspect ratios of the as-prepared ultralong hydroxyapatite nanowires were ultrahigh, which could reach more than 10,000, leading to high flexibility of ultralong hydroxyapatite nanowires, which can solve the problems of high brittleness and hardness of the hydroxyapatite materials. The author put forward the innovative idea of using highly flexible ultralong hydroxyapatite nanowires as the raw material for making a new type of fire-resistant inorganic paper. After many experiments, for the first time the author's team successfully developed a new type of fire-resistant inorganic paper.

2.3. SYNTHETIC METHODS, EQUIPMENT, AND CHARACTERIZATION TECHNIQUES

The synthesis of hydroxyapatite nanowires was reported in the literature. Several synthetic methods were reported for the synthesis of hydroxyapatite nanowires, including the solvothermal/hydrothermal method,[22–25] microwave-assisted synthesis,[26] hard template,[27, 28] sol-gel hydrothermal process,[29] and reverse micelles.[30] However, the lengths of hydroxyapatite nanowires were usually short and less than 10 μm, and these hydroxyapatite nanowires with relatively small aspect ratios were still brittle and not flexible. The synthesis of highly flexible ultralong hydroxyapatite nanowires with lengths larger than 100 μm and aspect ratios higher than 1,000 is a great challenge.[31, 32]

In 2013, the author's research group developed the calcium oleate precursor solvothermal method for the synthesis of ultralong hydroxyapatite nanowires.[12] Ultralong hydroxyapatite nanowires were synthesized using $CaCl_2$, oleic acid, $NaH_2PO_4 \cdot 2H_2O$, and NaOH in mixed solvents of water, ethanol and oleic acid by the calcium oleate precursor solvothermal method. The reaction suspension was transferred into a Teflon-lined stainless steel autoclave, sealed, and thermally treated at 180°C for different times (5–23 hours). Then, the reaction system was cooled down naturally to room temperature. The product after solvothermal treatment was collected, washed with ethanol, and deionized water several times, and dried at 60°C.

Caution: During the solvothermal/hydrothermal synthetic process, the temperature and pressure are high inside the autoclave, and there may be safety risks; the temperature and pressure inside the autoclave should be controlled strictly below the limits of the safe temperature and pressure.

Figure 2.2 shows the characterization results of ultralong hydroxyapatite nanowires synthesized by the calcium oleate precursor solvothermal method. The scanning electron microscopy (SEM) and transmission electron microscopy (TEM) images in Figure 2.2a–c show that the product consisted of single-crystalline ultralong hydroxyapatite nanowires. The as-prepared ultralong hydroxyapatite nanowires had diameters of ~10 nm and lengths of up to several hundred micrometers and very high aspect ratios (up to > 10,000). The as-prepared ultralong hydroxyapatite nanowires were highly flexible and could bend naturally. The X-ray diffraction (XRD) patterns in Figure 2.2e indicate that the precursor was a single phase of calcium oleate, and the product obtained by the calcium oleate precursor solvothermal method was single-phase hydroxyapatite. When ultralong hydroxyapatite nanowires were dispersed in ethanol and stirred with a glass rod, a long fiber with a length of 28 mm and a high flexibility was formed (Figure 2.2d). The high flexibility of ultralong hydroxyapatite nanowires was mainly attributed to their ultrahigh aspect ratios and ultralong lengths.

It was found that the hydrophilicity/hydrophobicity of ultralong hydroxyapatite nanowires could be adjusted by varying the experimental parameters, and both hydrophilic and hydrophobic ultralong hydroxyapatite nanowires could be prepared. The contact angles of the tablet samples prepared by pressing the as-prepared powders of ultralong hydroxyapatite nanowires were measured (Figure 2.2f–j). Ultralong hydroxyapatite nanowires synthesized using 0.0926 g of $NaH_2PO_4 \cdot 2H_2O$ for a solvothermal time of 5 hours was highly hydrophobic with a contact angle of 137°C (Figure 2.2f), but the contact angle decreased to 108°C for a solvothermal time of 23 hours (Figure 2.2g). When the amount of $NaH_2PO_4 \cdot 2H_2O$ was increased to 0.120 g, the contact angle of ultralong hydroxyapatite nanowires was 95°C and 20°C for a

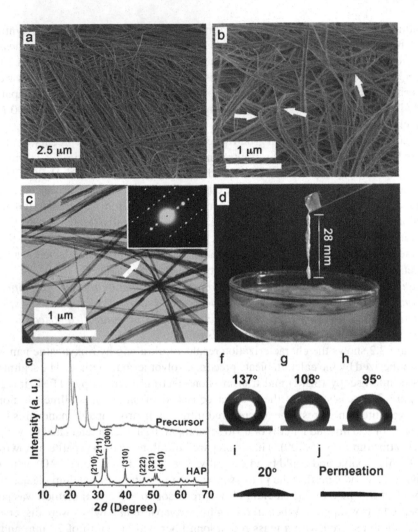

FIGURE 2.2 Characterization of ultralong hydroxyapatite nanowires prepared by the calcium oleate precursor solvothermal method. (a, b) Scanning electron microscopy (SEM) images; (c) transmission electron microscopy (TEM) image, and the inset of (c) was a selected-area electron diffraction pattern of a single hydroxyapatite nanowire; (d) the formation of a long fiber with a length of 28 mm obtained by stirring a dispersion of ultralong hydroxyapatite nanowires in ethanol; (e) X-ray diffraction (XRD) patterns of the calcium oleate precursor and ultralong hydroxyapatite nanowires (sample H-0.120 g–5 h); (f–j) water contact angle images of different samples: (f) H-0.0926 g–5 h, (g) H-0.0926 g–23 h, (h) H-0.120 g–5 h, (i) H-0.120 g–9 h, and (j) H-0.120 g–23 h; the samples synthesized with different amounts of $NaH_2PO_4 \cdot 2H_2O$ (labeled as x gram, for example, 0.120 g) and different solvothermal reaction times (labeled as y hours, for example, 5 hours), while the other experimental conditions were kept the same, were labeled as "H-x g–y h" (e. g., H-0.120 g–5 h). (Reprinted with permission from reference [12])

solvothermal time of 5 and 9 hours, respectively (Figure 2.2h and i), and ultralong hydroxyapatite nanowires with excellent hydrophilicity was obtained for a solvothermal time of 23 hours (Figure 2.2j). The difference in hydrophilicity/hydrophobicity was attributed to the adsorption of oleic acid molecules or oleate groups with long hydrophobic hydrocarbon chains on the surface of ultralong hydroxyapatite nanowires.

The calcium oleate precursor solvothermal method could be extended to the synthesis of ultralong hydroxyapatite nanowires using various monohydroxy alcohols[31] and a variety of phosphate salts.[32] For example, ultralong hydroxyapatite nanowires with lengths of close to 1 mm could be synthesized by the calcium oleate precursor solvothermal method using methanol instead of ethanol.[31] Figure 2.3 shows ultralong hydroxyapatite nanowires dispersed in deionized water after washing with ethanol and deionized water. From Figure 2.3 one can see that the as-prepared ultralong hydroxyapatite nanowires exhibited a high whiteness, good dispersibility, and high stability. The aqueous dispersion of ultralong hydroxyapatite nanowires was stable for a long period of time without obvious precipitation.

In addition, ultralong hydroxyapatite nanowires could also be rapidly synthesized by the highly efficient, low-cost, environmentally friendly and energy-saving microwave-assisted calcium oleate precursor solvothermal/hydrothermal method in a short period of heating time.[33] This method was based on the green chemistry

FIGURE 2.3 Ultralong hydroxyapatite nanowires dispersed in deionized water; ultralong hydroxyapatite nanowires were prepared by the calcium oleate precursor solvothermal method using methanol in a Teflon-lined stainless-steel autoclave with a volume of 1 L.

strategy, using water as the only solvent in the absence of organic solvents. Ultralong hydroxyapatite nanowires with diameters of tens of nanometers and lengths of hundreds of micrometers could be synthesized in a short period of time (within 20 minutes), which could significantly shorten the synthetic time by about two orders of magnitude compared with the conventional hydrothermal method, showing advantages such as high efficiency and significant energy saving.

A more environment-friendly and low-cost calcium oleate precursor hydrothermal method was developed for the synthesis of ultralong hydroxyapatite nanowires using water-soluble calcium salt such as $CaCl_2$, sodium oleate, and water-soluble phosphate such as NaH_2PO_4 in water as the only solvent without using any organic solvent.[34] The calcium oleate precursor hydrothermal method is different from the calcium oleate precursor solvothermal method for the synthesis of ultralong hydroxyapatite nanowires using water-soluble calcium salt such as $CaCl_2$, oleic acid, $NaH_2PO_4 \cdot 2H_2O$ (or other water-soluble phosphates), and NaOH in mixed solvents of water, ethanol (or other alcohols), and oleic acid. In the calcium oleate precursor hydrothermal method, monodisperse hydroxyapatite nanowires with lengths of several hundred nanometers are formed in the reaction system from the molecular to the nanoscale level, and then hydroxyapatite nanowires self-assemble into long fibers with lengths up to several hundred micrometers and further into two-dimensional nanowire networks from the nanoscale to the microscale. Monodisperse hydroxy-apatite nanowires were prepared by simply washing the hydrothermal product with a small amount of water to remove the impurities. Self-assembled hydroxyapatite nanowire networks were obtained by introducing ethanol into the hydrothermal product containing monodisperse hydroxyapatite nanowires.

Figure 2.4 shows the calcium oleate precursor hydrothermal synthesis and hierarchical assembly from monodisperse hydroxyapatite nanowires (Level 1), to long fibers (Level 2), and to nanowire networks (Level 3). As shown in Figure 2.4a, sodium oleate reacted with calcium chloride at room temperature to form calcium oleate as a precursor, which further reacted with NaH_2PO_4 to form monodisperse hydroxyapatite nanowires under hydrothermal conditions. Because of its small solubility product constant, calcium oleate could release Ca^{2+} ions slowly into the aqueous solution, leading to slow nucleation of hydroxyapatite and growth of nanowires. The oleate groups could be adsorbed on the surface of hydroxyapatite nanowires to form the oleate bilayer, leading to excellent monodispersity and self-assembly in parallel alignment of the as-prepared hydroxyapatite nanowires in the aqueous system (Figure 2.4b–d) due to the electrostatic interaction. The as-prepared hydroxyapatite nanowires had uniform diameters of 6.7 ± 1.0 nm and lengths of 436 ± 259 nm with an aspect ratio of about 65. The as-synthesized monodisperse hydroxyapatite nanowires could self-assemble into long fibers at the nanoscale by binding short nanowires in parallel arrangement induced by ethanol (Figure 2.4e and f), and further into densely connected two-dimensional nanowire networks by crosslinking long fibers at the microscale induced by ethanol (Figure 2.4g and h).

The experiments indicated that the oleate groups played a key role in the self-assembly of monodisperse hydroxyapatite nanowires into flexible two-dimensional nanowire networks. The formation of the bilayer of oleate groups adsorbed on the

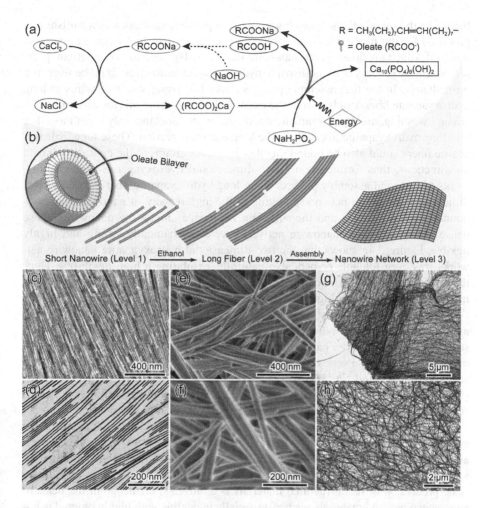

FIGURE 2.4 The calcium oleate precursor hydrothermal synthesis and hierarchical assembly from monodisperse hydroxyapatite nanowires (Level 1), to long fibers (Level 2), and to nanowire networks (Level 3). (a) Schematic illustration of the synthetic route to hydroxyapatite nanowires in an aqueous reaction system; (b) schematic illustration of the hierarchical assembly processes of hydroxyapatite nanowires to form hydroxyapatite nanowire networks; (c, d) TEM images of monodisperse hydroxyapatite nanowires; (e, f) SEM images of long fibers formed by self-assembly of hydroxyapatite nanowires; (g, h) TEM images of networked hydroxyapatite nanowires. (Reprinted with permission from reference [34])

surface of hydroxyapatite nanowires led to monodispersity of hydroxyapatite nanowires in aqueous solution. The paper sheet made from hydroxyapatite nanowires washed with ethanol twice exhibited a superhydrophilic property. However, the samples made from hydroxyapatite nanowires washed with ethanol twice and water once exhibited a high hydrophobicity (144°). This experimental result could be explained by that the bilayer of oleate groups was broken during the water washing process but not in the initial ethanol washing process. A possible formation mechanism was proposed for the hierarchical assembly from monodisperse hydroxyapatite nanowires to flexible two-dimensional nanowire networks. By the addition of ethanol, the self-assembly process of monodisperse hydroxyapatite nanowires could be triggered immediately. In the first assembly process (from hydroxyapatite nanowires to long hydroxyapatite fibers), only part of the outer shell of the oleate bilayer was broken by ethanol washing, and hydroxyapatite nanowires were spontaneously bound together into long hydroxyapatite fibers along the longitudinal direction. These long hydroxyapatite fibers could also be bound together by oleate groups at the contact points in all directions, thus forming dense two-dimensional hydroxyapatite nanowire networks (the second assembly process: from long hydroxyapatite fibers to dense two-dimensional nanowire networks). Owing to the high density of nanowire junctions bound by oleate groups and the sufficient interspace among the nanowires, these dense two-dimensional nanowire networks were mechanically strong and highly flexible. Figure 2.4g shows that the two-dimensional hydroxyapatite nanowire networks was even foldable or bendable naturally, exhibiting a high flexibility. Thus, the two-dimensional hydroxyapatite nanowire networks are an excellent building material for constructing highly flexible, high-strength and fire-resistant paper.[34]

Other compounds with similar structures and properties to sodium oleate, such as sodium stearate and sodium laurate, could also be used to synthesize network-structured hydroxyapatite nanowires by the calcium oleate precursor hydrothermal method. In addition, a series of commonly used water-soluble inorganic phosphates such as $(NaPO_3)_6$ and $Na_2HPO_4 \cdot 12H_2O$, $K_3PO_4 \cdot 3H_2O$, $K_2HPO_4 \cdot 3H_2O$, $K_4P_2O_7 \cdot 3H_2O$, $(NH_4)_3PO_4 \cdot 3H_2O$, $NH_4H_2PO_4$, and $(NH_4)_2HPO_4$ could be used as the phosphorous source for the synthesis of network-structured hydroxyapatite nanowires.

The calcium oleate precursor hydrothermal method could be scaled up for the synthesis of hydroxyapatite nanowires and nanowire networks using a stainless steel autoclave with a volume of 10 L.[34] Figure 2.5 shows the schematic illustration for the scaled-up synthesis of hydroxyapatite nanowires and nanowire networks. The large-scale production of nanostructured materials is a universal challenge in the fields of nanotechnology and materials science, especially in dealing with highly ordered nanostructured materials. However, the large-scale production of nanostructured materials is a key factor for the realization of their practical applications. The laboratory synthesis of nanostructured materials is usually in a small scale of less than 100 mL volume reaction system. The author's research group successfully realized large-scale (autoclaves with a volume of 10 L and 100 L) synthesis of hydroxyapatite nanowires and nanowire networks by the calcium oleate precursor hydrothermal method. This method is environmentally friendly, cost-effective, and thus more suitable for industrial production.[34]

The synthetic equipment for ultralong hydroxyapatite nanowires was the hydrothermal/solvothermal autoclave, the reaction liquid containing the reactants, additives, and solvents was added in the autoclave, sealed, and heated to a desired temperature and maintained at that temperature for a fixed time. *Caution: During the synthetic process, the temperature and pressure are high inside the autoclave, and there may be safety risks; the temperature and pressure inside the autoclave should be controlled strictly below the limits of the safe temperature and pressure.* In addition, the author's research

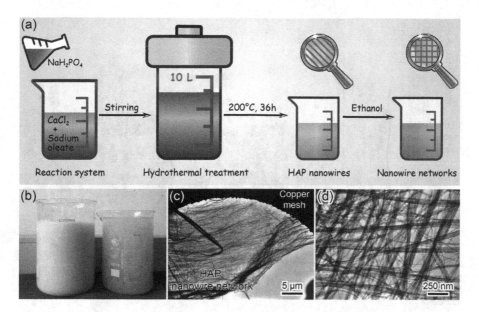

FIGURE 2.5 Scaled-up synthesis of hydroxyapatite nanowires/nanowire networks by the calcium oleate precursor hydrothermal method. (a) Schematic illustration of the scaled-up synthesis procedure; (b) digital images of the hydroxyapatite nanowires dispersed in aqueous solution (left beaker) and self-assembled nanowire networks in ethanol (right beaker); (c, d) TEM images of the as-prepared hydroxyapatite nanowire networks. (Reprinted with permission from reference [34])

group extended the synthesis to a larger scale, a larger stainless steel autoclave with a volume of 10 L and 100 L were used. Figure 2.6 shows various autoclaves used for the synthesis of ultralong hydroxyapatite nanowires in our laboratory.

As discussed above, in addition to SEM, TEM, XRD, and water contact angle, other commonly adopted characterization techniques for ultralong hydroxyapatite nanowires include Fourier transform infrared (FTIR) spectroscopy, surface area analysis, thermogravimetric and differential scanning calorimetry, zeta potential, etc.

2.4. PROPERTIES

As discussed above, ultralong hydroxyapatite nanowires synthesized by the calcium oleate precursor solvothermal method had diameters of about 10 nm and lengths of several hundred micrometers with ultrahigh aspect ratios of > 10,000. Ultralong hydroxyapatite nanowires exhibited high flexibility owing to their small diameters and ultrahigh aspect ratios. Highly flexible ultralong hydroxyapatite nanowires can overcome the high brittleness of traditional hydroxyapatite materials, and they can be used for making highly flexible functional materials.[35]

The hydrophilicity and hydrophobicity of the as-prepared ultralong hydroxyapatite nanowires could be controlled by adjusting synthetic parameters or post-washing process.[12] In addition to changing the experimental conditions, the post-washing

50 mL Teflon-
lined stainless
steel autoclave

1 L Teflon-lined   10 L stainless steel
stainless steel   autoclave
autoclave

100 L stainless steel autoclave

FIGURE 2.6 Teflon-lined stainless steel autoclaves with a volume of 50 mL and 1 L, respectively, stainless steel autoclaves with a volume of 10 L and 100 L, respectively.

process could also be used to control the surface properties of the as-prepared ultralong hydroxyapatite nanowires. Generally, thorough washing using ethanol and water many times of the product could produce hydrophilic ultralong hydroxyapatite nanowires; however, no washing or less washing would obtain hydrophobic ultralong hydroxyapatite nanowires. Therefore, depending on the purpose and application, ultralong hydroxyapatite nanowires with desirable surface properties could be prepared. Furthermore, the zeta potential measurements indicated that the as-prepared ultralong hydroxyapatite nanowires were negatively charged in aqueous solution.[34]

In addition, the as-prepared ultralong hydroxyapatite nanowires were stable in the weak acidic, neutral, and highly alkaline solutions, but were soluble in the strong acidic solution. This property could be used for some special applications such as the pH-responsive drug delivery.[36]

2.5. HIGHLY ORDERED SELF-ASSEMBLY AND DERIVED PRODUCTS

Currently, the preparation of large-sized highly ordered hydroxyapatite nanostructured materials remains a great challenge, especially for the fabrication of large-sized highly ordered ultralong hydroxyapatite nanowires because ultralong hydroxyapatite nanowires are easily tangled and aggregated. As discussed above, hydroxyapatite nanowires synthesized by the calcium oleate precursor hydrothermal method could self-assemble to form long fibers with an ordered alignment structure and further to form network-structured hydroxyapatite nanowires.[34]

The author's research group developed a novel strategy for the rapid automated preparation of highly flexible, large-sized, fire-resistant nanorope consisting of self-assembled highly ordered ultralong hydroxyapatite nanowires at room temperature, and the various derived flexible fire-resistant highly ordered architectures, such as highly flexible fire-resistant textiles, and three-dimensional printed well-defined highly ordered fire-resistant patterns. The nanorope was prepared by simply injecting the solvothermal product slurry containing ultralong hydroxyapatite nanowires into absolute ethanol at room temperature using the homemade automated equipment with round-end needles. The fire-resistant nanorope was successively formed from the nanoscale to the microscale then to the macroscale, and the ordering direction was controllable. The as-prepared fire-resistant nanorope was smooth on the surface and highly flexible and easy to be picked up and rolled up, as shown in Figure 2.7a and b. SEM images in Figure 2.7c–e show the highly ordered ultralong hydroxyapatite nanowires in the fire-resistant nanorope. The length of the fire-resistant nanorope could be controlled by injecting a certain amount of the solvothermal product slurry containing ultralong hydroxyapatite nanowires, and the diameter of the nanorope could be controlled by the inner diameter of the needle used for injection.[35]

A variety of biological materials in natural organisms provide excellent structural design guidelines and inspirations for the construction of advanced structural materials with excellent mechanical properties. Inspired by the natural nacre and human bone, the authors research group prepared a flexible macroscopic ribbon fiber made from highly ordered alignment of ultralong hydroxyapatite nanowires and sodium polyacrylate with a "brick-and-mortar" layered structure by a scalable and convenient injection method.

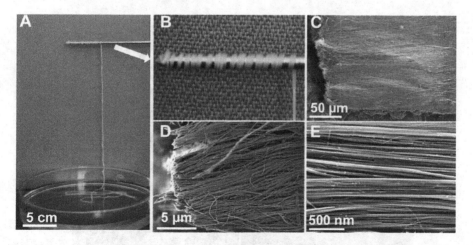

FIGURE 2.7 Characterization of the highly flexible, large-sized, fire-resistant nanorope consisting of self-assembled highly ordered ultralong hydroxyapatite nanowires prepared by simply injecting the solvothermal product slurry containing ultralong hydroxyapatite nanowires into absolute ethanol at room temperature. (A, B) Digital images showing the large-sized fire-resistant nanorope; (C–E) SEM images of the as-prepared fire-resistant nanorope consisting of highly ordered ultralong hydroxyapatite nanowires. (Reprinted with permission from reference [35])

The quasi-long-range orderly liquid crystal of ultralong hydroxyapatite nanowires was used and spun into the continuous flexible macroscopic ribbon fiber consisting of highly ordered ultralong hydroxyapatite nanowires. The highly ordered ultralong hydroxyapatite nanowires acted as the hard "brick" and sodium polyacrylate acted as the soft "mortar," and the nacre-mimetic layer-structured architecture was obtained. The as-prepared flexible macroscopic ribbon fiber with a highly ordered structure showed superior mechanical properties, and the maximum tensile strength and Young's modulus were as high as 203.58 ± 45.38 MPa and 24.56 ± 5.35 GPa, respectively. In addition, benefiting from the excellent flexibility and good knittability, the as-prepared macroscopic ribbon fiber could be woven into various flexible macroscopic architectures. Additionally, the as-prepared flexible macroscopic ribbon fiber could be further functionalized by incorporation of various functional components, such as magnetic and photoluminescent constituents. The as-prepared flexible macroscopic ribbon fiber is promising for applications in various fields such as smart wearable devices, optical devices, magnetic devices, and biomedical engineering.[37]

2.6. COMPARISON OF ULTRALONG HYDROXYAPATITE NANOWIRES WITH CELLULOSE FIBERS

The main differences between ultralong hydroxyapatite nanowires and cellulose fibers are summarized as follows:

(1) *Classification*: hydroxyapatite is inorganic; however, cellulose is organic.

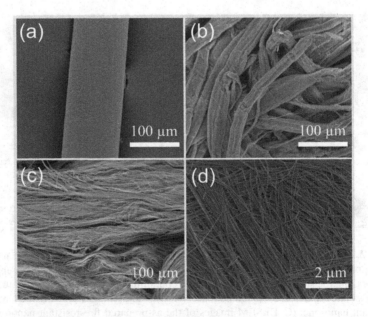

FIGURE 2.8 Size and morphology comparison of a human hair (a), cellulose fibers (b), and ultralong hydroxyapatite nanowires (c, d).

(2) *Chemical composition*: cellulose $(C_6H_{10}O_5)_n$ is a long-chain polymeric polysaccharide carbohydrate consisting of glucose units and constitutes the main part of the cell walls of plants, and is the raw material of paper. Hydroxyapatite $(Ca_{10}(OH)_2(PO_4)_6)$ is also called "basic calcium phosphate" or hydroxyl calcium phosphate, and is a member of the calcium phosphate family, which is the main inorganic constituent of bone and tooth.

(3) *Size and morphology*: cellulose fibers have a belt morphology with widths in the microscale and lengths in the millimeter or centimeter scale, depending on the type of the plant. In contrast, ultralong hydroxyapatite nanowires have diameters of about 10 nm and lengths of several hundred micrometers. Figure 2.8 shows the size and morphology comparison of ultralong hydroxyapatite nanowires with cellulose fibers and a human hair.

(4) *Flexibility*: both cellulose fibers and ultralong hydroxyapatite nanowires have good flexibility.

(5) *Thermal stability*: cellulose fibers are thermally stable below about 200°C. Cellulose fibers will be carbonized, blackened, and even burned to ashes at higher temperatures. However, ultralong hydroxyapatite nanowires have very high thermal stability; they are still stable at temperatures above 1,000°C.

(6) *Fire resistance*: cellulose fibers are highly flammable; that is why the traditional paper can be burned to ashes in seconds in fire. In contrast, ultralong hydroxyapatite nanowires are nonflammable and have excellent fire-resistant performance.

3 Novel Fire-Resistant Paper Based on Ultralong Hydroxyapatite Nanowires

3.1. BACKGROUND FOR THE INVENTION OF THE FIRE-RESISTANT PAPER

As discussed earlier, although hydroxyapatite has excellent biocompatibility and environmental friendliness, high whiteness, high melting point (~1650°C), high-temperature resistance, and nonflammability, unfortunately, hydroxyapatite is usually as hard and brittle as tooth and bone, and is not suitable for making soft fire-resistant paper.

Since 2008, the author's research team has done a lot of research work on hydroxyapatite nanostructured materials and their applications including drug delivery,[36, 38–51] bio-imaging,[40] and bone defect repair.[45, 52–54] In 2013, the author's research group developed the calcium oleate precursor solvothermal method, and successfully synthesized ultralong hydroxyapatite nanowires by this method. The lengths of the as-prepared ultralong hydroxyapatite nanowires were several hundred micrometers, and the diameters of nanowires were only about 10 nm, which is equivalent to about 1/10,000 of the thickness of human hair. The aspect ratios of the as-prepared ultralong hydroxyapatite nanowires were ultrahigh, which could reach more than 10,000, leading to high flexibility of ultralong hydroxyapatite nanowires, which can solve the problem of high brittleness and hardness of traditional hydroxyapatite materials. The author put forward the innovative idea of using highly flexible ultralong hydroxyapatite nanowires as the raw material to develop a new kind of fire-resistant paper. After many experiments, the author's team successfully developed a new type of fire-resistant paper based on ultralong hydroxyapatite nanowires for the first time.[12] The fire-resistant paper based on ultralong hydroxyapatite nanowires is in a real sense a new kind of inorganic fire-resistant paper with high flexibility, high whiteness, high surface smoothness, good mechanical properties, excellent resistance to both fire and high temperatures, and high similarity to traditional cellulose fiber paper in appearance. In addition, the fire-resistant paper based on ultralong hydroxyapatite nanowires can be used for writing and printing, which is the basic function of paper. Importantly, the fire-resistant paper based on ultralong hydroxyapatite nanowires exhibits superior mechanical properties because of strong interactions such as hydrogen bonding between ultralong hydroxyapatite nanowires.

Two-dimensional fire-resistant products, including fire-resistant cloths, felts, and blankets, are important energy-saving materials in high-temperature industrial fields. Inorganic fibers are common building materials to make fire-resistant paper-like products. Common types of inorganic fibers include aluminum silicate, magnesium silicate, basalt, aluminum oxide, silicon dioxide, etc. Inorganic fibers are commonly produced by the wire-drawing process in the melting state. However, inorganic fibers produced by common methods have usually large diameters (usually more than 10 μm) and low aspect ratios, with poor flexibility. The paper sheets made from these thick inorganic fibers usually have poor mechanical properties due to the weak interactions between thick fibers (no hydrogen bonds can be formed between many inorganic fibers because they contain no hydroxyl groups or other functional groups which can form hydrogen bonds), and have very rough surface which cannot be used for writing and printing and can only be used for industrial purposes. In addition, since paper is frequently used in daily work and life, the biocompatibility of paper is an important issue. Although some inorganic fibers, such as asbestos fibers, are used to make special paper-like products, the asbestos fibers are toxic and have a major concern to human health.[55]

3.2. PREPARATION OF THE FIRE-RESISTANT PAPER

3.2.1. RAW MATERIALS AND STRUCTURAL DESIGN OF THE FIRE-RESISTANT PAPER

The single-phase fire-resistant paper consisting of only ultralong hydroxyapatite nanowires as the raw material without any additives exhibited poor mechanical properties—for example, its tensile strength was usually lower than 1 MPa. In order to enhance the mechanical properties of the fire-resistant paper, the author's research group designed and optimized the structure of the fire-resistant paper. The new type of fire-resistant paper possessed a "steel-reinforced concrete structure," similar to that used in tall buildings; it consisted only of inorganic components, including ultralong hydroxyapatite nanowires as the main building material (like a cement), micrometer-sized inorganic fibers as the reinforcing skeleton material (like concrete reinforcing steel bars), and an inorganic adhesive as the binder. In the fire-resistant paper, ultralong hydroxyapatite nanowires were mutually interwoven to form a porous network structure, coupled with the mechanical support and reinforcement of the micrometer-sized inorganic fibers and the binding effect of the inorganic adhesive; the three synergistic effects could significantly enhance the mechanical properties of the fire-resistant paper.

3.2.2. PAPERMAKING PROCESSES AND EQUIPMENT

The new kind of fire-resistant paper is prepared using ultralong hydroxyapatite nanowires as the main raw material, a certain amount of micrometer-sized inorganic fibers such as glass fibers, and an inorganic adhesive prepared in the author's laboratory. The preparation process of the fire-resistant paper is similar to that of traditional cellulose fiber paper, including the following main procedures: (1) the

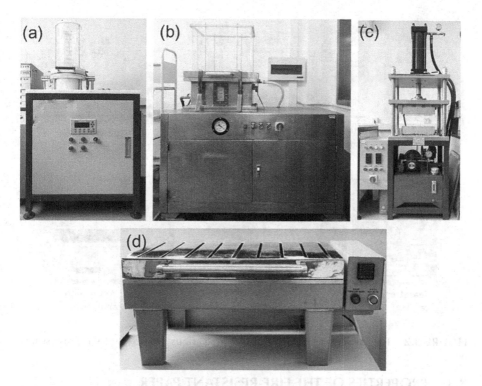

FIGURE 3.1 Papermaking equipment for the preparation of the fire-resistant paper. (a) A paper sheet former with a diameter of 20 cm; (b) an A3-size (420 mm × 297 mm) paper sheet former; (c) a presser with a heating function and temperature controller; (d) a paper dryer with a flat surface.

preparation of aqueous suspension containing ultralong hydroxyapatite nanowires, micrometer-sized inorganic fibers, and inorganic adhesive; (2) the vacuum-assisted filtration of the aqueous suspension; (3) pressing; and (4) drying. It should be noted that traditional papermaking needs bleaching and sizing procedures; however, bleaching and sizing are not necessary for the preparation of the fire-resistant paper. Therefore, the papermaking process for fire-resistant paper is more environmentally friendly and cost-effective compared with traditional papermaking. In addition, traditional papermaking equipment was used for the preparation of the fire-resistant paper in the author's laboratory, as shown in Figure 3.1.

3.2.3. CHARACTERIZATION TECHNIQUES

The properties of the fire-resistant paper can be measured and analyzed using various instruments used for traditional paper based on plant fibers. These instruments include the thickness tester, whiteness tester, smoothness tester, folding resistance tester, stiffness tester, tear strength tester, bursting strength tester, water absorption tester, tensile tester, and water content analyzer, as shown in Figure 3.2.

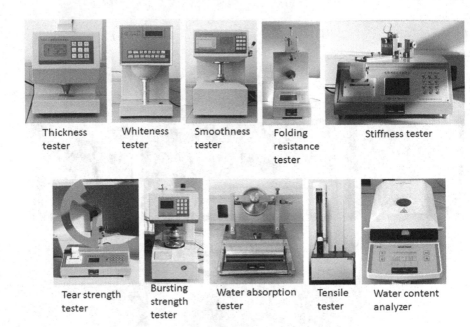

FIGURE 3.2 The instruments used for measuring various properties of the fire-resistant paper.

3.3. PROPERTIES OF THE FIRE-RESISTANT PAPER

The new type of fire-resistant paper has high flexibility and can be curled arbitrarily. It shows high whiteness without adding any bleaching agent, which is environment friendly, unlike traditional papermaking in which bleaching is necessary. It should be noted that one unique feature of the fire-resistant paper is its layered structure in the thickness direction,[56, 57] which is different from traditional paper based on plant fibers. The fire-resistant paper can be used for writing and color printing, which is the primary function of paper. Figure 3.3 shows digital images of an A4-sized (29.7 cm × 21 cm) fire-resistant paper consisting of ultralong hydroxyapatite nanowires as the main raw material. The as-prepared A4-sized highly flexible fire-resistant paper exhibited a high whiteness, and it looked like traditional cellulose fiber paper in appearance. The fire-resistant paper could be bent and rolled without any visible damage, indicating the high flexibility of this new kind of fire-resistant inorganic paper. The text could be clearly printed in color on the fire-resistant paper by directly using a commercial ink-jet printer.

The most amazing thing is that the as-prepared ultralong hydroxyapatite nanowires-based highly flexible inorganic paper is highly resistant to both high temperature and fire. Figure 3.4a shows the excellent fire-resistant performance of the ultralong hydroxyapatite nanowires-based inorganic paper. The temperature in the flame of an alcohol lamp is close to 700°C; the fire-resistant paper could not be burned in the flame even after a long period of time, exhibiting an excellent fire-resistant performance. In contrast, traditional cellulose fiber paper was immediately burned to

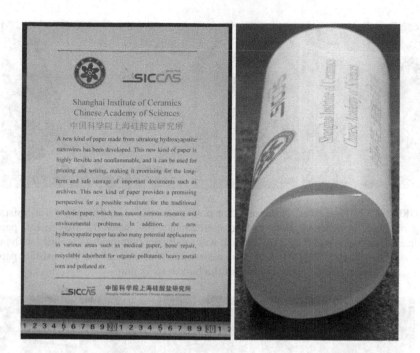

FIGURE 3.3 Digital images of an A4-size fire-resistant paper consisting of ultralong hydroxyapatite nanowires as the main raw material. The A4-size fire-resistant paper could be directly used for color printing using a commercial ink-jet printer. (Reprinted with permission from reference [34])

ashes in just a few seconds in the flame of an alcohol lamp. It is expected that the fire-resistant paper can be used for permanent, safe preservation of important documents, archives, and books, avoiding their destruction in fire. In addition, the butane spray gun was also used to test the resistance performance to high temperature and fire. Although the flame temperature of the butane spray gun was about 1300°C, the fire-resistant paper could well resist such a high temperature, and was not flammable at all, as shown in Figure 3.4b.

The fire-resistant paper based on ultralong hydroxyapatite nanowires exhibits an excellent thermal insulation performance. Figure 3.5 shows the adiabatic and flexible properties of the fire-resistant paper under high-temperature conditions. As shown in Figure 3.5a and b, the common filter paper based on cellulose fibers and the fire-resistant paper were placed on the flame of a spirit lamp with a piece of silk cocoon placed on them. When being exposed to the flame, the common filter paper and the silk cocoon immediately caught fire and were burned to ashes in only 5 seconds. However, the fire-resistant paper with written characters and the silk cocoon on it could be well preserved after being exposed to the flame for 5 minutes, indicating the excellent fire resistance and heat insulation properties of the fire-resistant paper. Figure 3.5c shows the temperature variation curves of a thermocouple wrapped with a 150 μm-thick fire-resistant paper and a bare thermocouple which were placed on

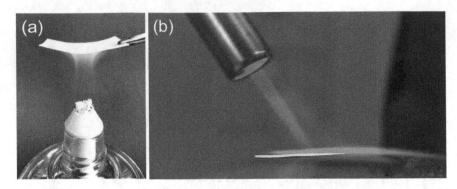

FIGURE 3.4 (a) The excellent fire-resistant performance of the as-prepared fire-resistant inorganic paper based on ultralong hydroxyapatite nanowires; when the fire-resistant paper was heated in the flame of the alcohol lamp, it was totally nonflammable (Reprinted with permission from reference [58]); (b) the excellent resistant performance of the fire-resistant paper to high flame temperature (about 1300°C) of the butane spray gun.

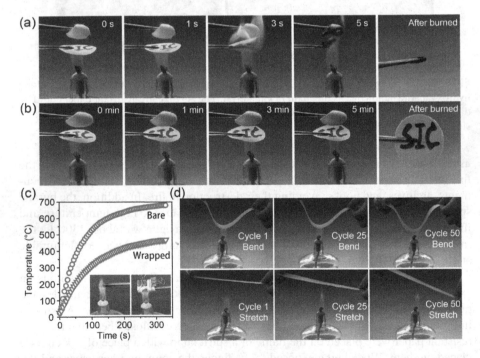

FIGURE 3.5 Adiabatic and flexible properties of the fire-resistant paper based on ultralong hydroxyapatite nanowires under high-temperature conditions. (a, b) Thermal insulation tests: protecting silk cocoon from fire using the common filter paper (a) and fire-resistant paper (b); (c) temperature variation curves of a bare thermocouple and 150 μm-thick fire-resistant paper wrapped thermocouple placed on the flame; (d) flexibility tests of the fire-resistant paper by performing bending-stretching cycles on the flame. (Reprinted with permission from reference[34])

the flame of a spirit lamp. After being heated for more than 5 minutes, the temperature of the thermocouple gradually reached an equilibrium, and the temperature of the thermocouple wrapped with the fire-resistant paper was more than 200°C lower than that of the bare thermocouple, showing excellent adiabatic property and the potential heat-shielding application of this new kind of fire-resistant paper. In addition, the fire-resistant paper could maintain a good flexibility under high-temperature conditions, as shown in Figure 3.5d. The fire-resistant paper could be bent and stretched back and forth on the flame for more than 50 cycles without breaking, indicating the good mechanical properties and flexibility even under high-temperature conditions.[34]

Furthermore, ultralong hydroxyapatite nanowires have a pH-responsive property. Ultralong hydroxyapatite nanowires are stable in weak acidic, neutral, and strong alkaline solutions. However, ultralong hydroxyapatite nanowires will be rapidly dissolved in a strong acidic solution. This property can be used to destroy the ultralong hydroxyapatite nanowires-based fire-resistant paper when necessary.

3.4. ADVANTAGES OF THE FIRE-RESISTANT PAPER

The fire-resistant paper based on ultralong hydroxyapatite nanowires has obvious advantages, for example, (1) high flexibility; (2) high-quality white color; (3) high biocompatibility and environmental friendliness; (4) no need for bleaching and sizing, which are required for traditional cellulose fiber paper; (5) can be used for writing and color printing using a commercial printer; (6) highly resistant to high temperatures; (7) excellent fire resistance; (8) can be used for long-term safe preservation of important documents such as archives and books; (9) has many other applications; and (10) good application prospects in many fields. In addition, the ultralong hydroxyapatite nanowires, which are the raw material of the new type of fire-resistant paper, can be synthesized artificially using common chemicals without consuming valuable natural resources such as trees and without negatively impacting the environment. The whole manufacturing process of the fire-resistant paper is environmentally friendly and will not pollute the environment; thus, the new fire-resistant paper has a tempting prospect for commercialization and large-scale applications. It is expected that in the near future the new fire-resistant paper may be able to move from the laboratory to the market, into book stores and libraries, to protect important documents, archives, and books, and for applications in various fields.

4 Functionalized Fire-Resistant Paper

4.1. INTRODUCTION TO FUNCTIONALIZATION OF FIRE-RESISTANT PAPER

The fire-resistant paper based on ultralong hydroxyapatite nanowires has the same functions as traditional paper based on plant fibers, such as writing and printing. In addition, the fire-resistant paper has abilities such as excellent resistance to both high temperatures and fire, which traditional paper does not possess. Therefore, the fire-resistant paper can be used in a variety of fields where traditional paper cannot be applied. In order to expand the application areas, the functionalization of the fire-resistant paper is a useful strategy.

Up until now, the author's research group has successfully developed more than 20 new types of fire-resistant paper based on ultralong hydroxyapatite nanowires with various functions, for example, the waterproof fire-resistant paper, fire-resistant "Xuan paper," high-temperature-resistant label paper, antibacterial fire-resistant paper, waterproof photoluminescent fire-resistant paper, waterproof electrically conductive fire-resistant paper, catalytic fire-resistant paper, waterproof magnetic fire-resistant paper, automatic fire-alarm fire-resistant wallpaper, photothermal fire-resistant paper, high-temperature-resistant battery separator, fire-resistant paper tape for electric cables and fiber-optic cables, secret information encryption and decryption fire-resistant paper, light-driven self-propelled waterproof fire-resistant paper, water purification filter paper, air purification filter paper, biomedical paper, antibacterial biomedical paper, rapid test paper, etc. In the following sections, some representative examples of the functionalized fire-resistant papers will be discussed.

4.2. WATERPROOF FIRE-RESISTANT PAPER

As the saying goes, "water and fire are ruthless." Both water and fire are natural enemies of paper. Traditional paper made from plant fibers is neither fire-resistant nor waterproof. These shortcomings limit the applications of traditional paper, and the long-term preservation of valuable documents, archives, and books has a potential danger. For many centuries, countless precious paper cultural relics, documents, and books were destroyed in fires and floods. For example, in January 2014, the French National Library suffered a flood, and more than 10,000 precious books were damaged, and the same tragedy struck the library again in 2004. On January 30, 2015, a major fire broke out at the Institute for Scientific Information on Social Sciences Library in Moscow and lasted for more than ten hours, and millions of precious ancient documents, books, and other materials were ruined in the fire. It is even worse that more documents and books were damaged during the fire extinguishing process, when a large amount of water was used. Is it possible to develop a super paper that is both waterproof and fire-resistant? If you have this kind of super paper that is not afraid of water and fire, the tragedy of a large number of precious paper cultural relics and books being destroyed in fire and floods could be avoided. As discussed above, the new type of fire-resistant paper we developed was not afraid of fire, it was still afraid of water, so it was necessary to tackle the non-waterproof problem of the fire-resistant paper.

To solve the non-waterproof problem of the new type of fire-resistant paper based on ultralong hydroxyapatite nanowires, it is necessary to make the fire-resistant paper have the superhydrophobic property (superhydrophobicity). What is superhydrophobicity? Superhydrophobicity means that the stable contact angle of water droplets on the surface of the material is greater than 150° and the rolling contact angle is less than 10°; materials with the superhydrophobic property have the advantages of being waterproof, anti-fouling, anti-fog, self-cleaning, etc., and have good application prospects in many fields.

The superhydrophobicity and self-cleaning functions of the lotus leaf, namely the lotus leaf effect, have aroused great interest in researchers. The construction of superhydrophobic materials usually imitates the surface structure of the lotus leaf. Figure 4.1 shows a photograph of spherical water droplets on a lotus leaf. There are two surface structural features in the lotus leaf, one is the unique micro-nano-sized structures; the other is a waxy layer consisting of a low-surface-energy substance on the surface. The surface of the lotus leaf has many micrometer-sized convex structures, and many nanostructures grow on the surface of each micrometer-sized convexity, thus forming many micro-nano-sized small chambers filled with air. Water droplets will form spherical water droplets on the surface of the lotus leaf due to the effect of surface tension, because the surface tension of water is the smallest and most stable in the spherical state. Spherical water droplets usually have sizes in the millimeter scale, and cannot enter smaller chambers filled with air, and they can

FIGURE 4.1 A photograph showing spherical water droplets on a lotus leaf. (Courtesy: http://www.ivsky.com/tupian/heye_v23651/)

only roll around on the top of smaller air chambers, that is, these small air chambers act as physical supports for spherical water droplets. On the other hand, the surface of the lotus leaf is also covered with a layer of a biological wax-like substance with a low surface energy. The micro-nano-sized structures and the low-surface-energy substance on the surface of the lotus leaf work together, resulting in the superhydrophobic property. The contact area between the surface of the lotus leaf and a water droplet or dust is small, thus spherical water droplets on the surface of the lotus leaf can take away the dust; that is, the lotus leaf exhibits the self-cleaning function.

In recent years, superhydrophobic materials have caught the attention of researchers owing to their high application value. Researchers used various methods to prepare superhydrophobic materials with regard to two aspects: (1) constructing micro-nano-sized structures on the surface of materials, and (2) reducing the surface energy. However, some preparation methods need strict requirements for equipment, leading to high cost; some methods use fluorine-containing reagents to chemically modify the surface of the materials to reduce its surface energy. However, fluorine-containing compounds are generally expensive and have certain toxicity, and have safety concerns in terms of the human health and the environment.

In 2016, the author's research group successfully developed a new kind of waterproof fire-resistant paper made from ultralong hydroxyapatite nanowires with adsorbed oleic acid molecules on the surface.[59] The ultralong hydroxyapatite nanowires with oleic acid molecules adsorbed on the surface can be artificially synthesized using common chemicals by the calcium oleate precursor solvothermal method[12] or the calcium oleate precursor hydrothermal method.[34] For example, ultralong hydroxyapatite nanowires could be synthesized in a reaction system containing $CaCl_2$, NaOH, oleic acid, NaH_2PO_4, ethanol, and deionized water by the calcium oleate precursor solvothermal method at 180°C in 24 hours. The as-prepared ultralong hydroxyapatite nanowires were stirred in ethanol and water, separately, at 60°C overnight to remove the impurities, and stored in deionized water. A certain volume of the colloid suspension containing ultralong hydroxyapatite nanowires was treated by vacuum filtration to form a wet paper sheet, which was immersed in a sodium oleate aqueous solution with a concentration of 0.02 mol L^{-1} for 1 hour under magnetic agitation, followed by vacuum-assisted filtration. The waterproof fire-resistant paper was obtained after drying at 60°C.[59]

Ultralong hydroxyapatite nanowires had no obvious change after surface modification with sodium oleate, as shown by the XRD patterns in Figure 4.2. FTIR spectra in Figure 4.3 of ultralong hydroxyapatite nanowires before and after surface modification were consistent with the XRD analysis. The FTIR spectra of ultralong hydroxyapatite nanowires before and after surface modification exhibited similar characteristic peaks of the hydroxyl group (3,565 and 633 cm^{-1}), PO_4^{3-} (1,093, 1,028, 962, 604, and 561 cm^{-1}), and adsorbed water (3,442 and 1,635 cm^{-1}). The peaks at 2,921 and 2,852 cm^{-1} in the FTIR spectra of ultralong hydroxyapatite nanowires after surface modification were attributed to the asymmetric and symmetric C–H stretching vibration of the alkyl group of oleate.

In fact, a simpler method was developed by the author's research group to prepare the waterproof fire-resistant paper. Oleic acid molecules were adsorbed on the

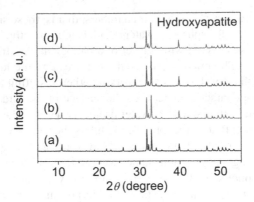

FIGURE 4.2 XRD patterns of ultralong hydroxyapatite nanowires modified with 0.02 mol L^{-1} sodium oleate aqueous solution for different times: (a) without surface modification; (b) 1 hour; (c) 2 hours; (d) 3 hours. (Reprinted with permission from reference [59])

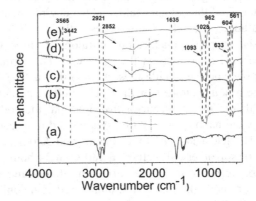

FIGURE 4.3 FTIR spectra: (a) sodium oleate; (b–e) ultralong hydroxyapatite nanowires modified with 0.02 mol L^{-1} sodium oleate aqueous solution for different times: (b) without surface modification; (c) 1 hour; (d) 2 hours; (e) 3 hours. (Reprinted with permission from reference [59])

surface of ultralong hydroxyapatite nanowires synthesized by the calcium oleate precursor solvothermal method or the calcium oleate precursor hydrothermal method. If the as-prepared ultralong hydroxyapatite nanowires were slightly washed using ethanol once, and collected and dispersed in water, then the waterproof fire-resistant paper could be obtained using the ultralong hydroxyapatite nanowires. If ultralong hydroxyapatite nanowires were thoroughly washed using ethanol and water for many times, the fire-resistant paper obtained was superhydrophilic instead of superhydrophobic. In addition, the entire manufacturing process of the new waterproof fire-resistant paper is environmentally friendly without causing severe pollution to the environment.

Why is oleic acid used in the preparation of the waterproof fire-resistant paper? Oleic acid has many advantages, such as non-toxicity, environment friendliness, and relatively low price. Oleic acid is widely present in nature, mainly in the form of glyceride in animal and vegetable oils. Oleic acid plays an important role in the metabolism of the human body, and is an indispensable nutrient in food. To cite a few examples of vegetable oils, the content of oleic acid is as high as 80% or more in tea oil, more than 50% in peanut oil, and 55–83% in olive oil. In addition, the sodium or potassium salt of oleic acid is one of the ingredients of soap.

How is the superhydrophobicity of the waterproof fire-resistant paper realized? In the process of making the waterproof fire-resistant paper, ultralong hydroxyapatite nanowires with oleic acid molecules adsorbed on their surface form a porous networked structure through overlapping, interweaving, and entanglement. This provides the paper with unique micro-nano-sized structures; in addition, the oleic acid molecules adsorbed on the surface of ultralong hydroxyapatite nanowires are a low-surface-energy substance. The synergy of these two combined features results in the superhydrophobic effect with the waterproof function, as shown in Figure 4.4. The waterproof fire-resistant paper has a variety of advantages, such as high flexibility, environmental friendliness, excellent and stable superhydrophobic performance, good self-cleaning function, excellent heat insulation, and fire resistance. The experiments showed that the waterproof fire-resistant paper exhibited excellent superhydrophobic property not only for water but also for various commercial beverages, such as mineral water, orange juice, tea, milk, and coffee. In addition to excellent superhydrophobic property, the waterproof fire-resistant paper also possessed excellent resistance to fire.

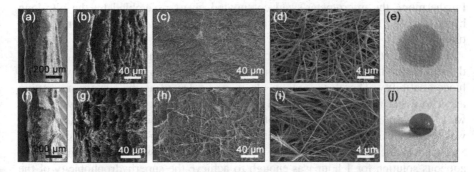

FIGURE 4.4 (a, b) SEM images of the cross-section of the fire-resistant paper made from ultralong hydroxyapatite nanowires synthesized by the calcium oleate precursor solvothermal method and washed thoroughly with ethanol and deionized water, that is, clean ultralong hydroxyapatite nanowires with almost no oleic acid molecules adsorbed on the surface (unmodified), showing a layered structure; (c, d) SEM images of the surface morphology of the unmodified fire-resistant paper; (e) a water droplet on the unmodified fire-resistant paper; (f, g) SEM images of the cross-section of the waterproof fire-resistant paper with surface modification using sodium oleate aqueous solution, showing a layered structure as well; (h, i) SEM images of the surface morphology of the layer-structured waterproof fire-resistant paper with surface modification; (j) a water droplet on the layer-structured waterproof fire-resistant paper with surface modification. (Reprinted with permission from reference [59])

The well-dispersed ultralong hydroxyapatite nanowires after thorough washing could be dispersed in water to form a stable wool-like suspension. After vacuum-assisted filtration, the ultralong hydroxyapatite nanowires self-assembled into a free-standing, layer-structured, highly flexible fire-resistant paper (Figure 4.4a and b). The layered structure can be clearly observed in Figure 4.4b. SEM images (Figure 4.4c and d) show that the ultralong hydroxyapatite nanowires in the paper were inter-twined with one another to form a porous network. The ultralong hydroxyapatite nanowires self-assembled along the longitudinal direction into nanowire bundles to form nanofibers and microfibers. Such a hierarchical structure could increase the roughness of the paper surface and thus achieve an excellent superhydrophobic per-formance. Without surface modification of the ultralong hydroxyapatite nanowires after thorough washing, the paper exhibited an intrinsic superhydrophilicity, and a water droplet could spread out immediately when it was dropped on the paper sur-face (water contact angle was ~0°) (Figure 4.4e). In the process of vacuum-assisted filtration, a layered structure was spontaneously formed (Figure 4.4a and b). During the filtration, ultralong hydroxyapatite nanowires were inclined to adhere together by the van der Waals force and hydrogen bonding between oxygen (hydrogen)-contain-ing groups. In addition, the electrostatic repulsion forces induced by the negatively charged ultralong hydroxyapatite nanowires could prevent them from complete compacting. Furthermore, the interaction among water molecules adsorbed on the oxygen-containing groups of the ultralong hydroxyapatite nanowires led to repul-sive hydration forces. The balance between the attractive forces (van der Waals and hydrogen bonding) and repulsive forces (electrostatic repulsive and hydration forces) was responsible for the formation of the layer-structured fire-resistant paper. Furthermore, the micrometer-sized corrugated layers assembled by the ultralong hydroxyapatite nanowires also contributed to the formation of the layered struc-ture.[57, 59]

Compared with expensive perfluorinated compounds, the sodium oleate used for the surface modification of ultralong hydroxyapatite nanowires is low-cost with good biocompatibility. Therefore, sodium oleate was selected as the surface-modifying agent to lower the surface energy of the ultralong hydroxyapatite nanowires. The waterproof fire-resistant paper exhibited a superhydrophobicity, and a water droplet was nearly spherical on its surface with a water contact angle of >150° (Figure 4.4j). Surface modification of the ultralong hydroxyapatite nanowires with sodium oleate aqueous solution for 1 hour was enough to achieve the superhydrophobicity of the fire-resistant paper (contact angle = $154.40 \pm 1.05°$). The SEM images in Figure 4.4f–i show that the waterproof fire-resistant paper made from surface-modified ultralong hydroxyapatite nanowires exhibited similar surface and cross-sectional morpholo-gies to those of the superhydrophilic fire-resistant paper made from unmodified ultralong hydroxyapatite nanowires. The rough hierarchical surface structure and the adsorbed oleic acid molecules as the low-surface-energy compound allowed the fire-resistant paper to have the superhydrophobic property. No color trace was left on the waterproof fire-resistant paper after it was dipped in and then taken out from water dyed with methylene blue, as shown in Figure 4.5a, exhibiting the excellent superhydrophobic property of the waterproof fire-resistant paper.

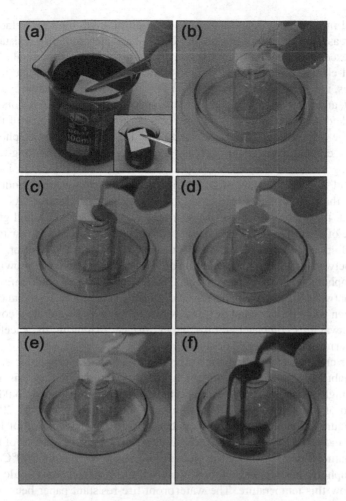

FIGURE 4.5 Liquid repellency tests for the waterproof fire-resistant paper. (a) A piece of the waterproof fire-resistant paper was immersed in deionized water (dyed with methylene blue) and taken out from the water (inset); (b–f) five common commercial drinks, including mineral water (b), red tea (c), orange juice (d), milk (e), and coffee (f), were poured on the waterproof fire-resistant paper. The waterproof fire-resistant paper exhibited excellent waterproof performance. After immersing into the dyed water and then taking it out, the waterproof fire-resistant paper was still white as before, without any color or water pollution on the paper. (Reprinted with permission from reference [59])

Liquid repellency performance of the waterproof fire-resistant paper was tested using pure water and various commercial drinks. As shown in Figure 4.5b–f, five common commercial drinks were poured onto the waterproof fire-resistant paper surface. In all experiments, not only transparent liquids (mineral water and red tea) but also suspensions and emulsions (orange juice, milk, and coffee) could easily roll off the horizontal surface of the waterproof fire-resistant paper. There were abundant nanopores on the waterproof fire-resistant paper surface (Figure 4.4h and i), and the air trapped in nanopores would lower the fraction of solid–liquid interface, which could increase the water contact angle. Thus, the waterproof fire-resistant paper exhibited excellent liquid repellency performance to various commercial drinks.[59]

The self-cleaning effect, one of the most attractive properties of superhydrophobic surfaces, is an effective tool to prevent the surfaces from the contamination with water, dust, and other pollutants. The contaminants on the superhydrophobic surface can be picked up and removed by water, such as rain. As an example of the application of the self-cleaning effect, the shading of the collector area of photovoltaic modules by contaminants could decrease the power output in the long-term. The self-cleaning ability of the superhydrophobic surface may provide an ideal strategy to avoid the accumulation of contaminants on the photovoltaic modules, thus improving their performance and extending their working lifetime. As shown in Figure 4.6a, the waterproof fire-resistant paper was attached to a tilted glass slide at an angle of about 15°. The small nanopores in the waterproof fire-resistant paper were small enough to prevent the dirt from penetrating into the interior. The contact area between the paper surface and water was reduced drastically owing to the superhydrophobicity of the paper surface. The tiny contact area could minimize the adhesion between dirt and the paper surface. Both effects resulted in enhanced adhesion between water droplets and dirt; thus water droplets and water flow could carry and wash away dirt completely on the paper surface, exhibiting the excellent self-cleaning performance (Figure 4.6c–f).[59]

Another challenge for the practical application of superhydrophobic surfaces is thermal stability. The superhydrophobicity of the waterproof fire-resistant paper exhibited high thermal stability, its superhydrophobic performance could be well maintained at a temperature of 100 °C for 24 hours (Figure 4.7a) or 200 °C for 1 hour (Figure 4.7b). The thermogravimetric analysis of the waterproof fire-resistant paper indicated that the oleate molecules adsorbed on the surface of ultralong hydroxyapatite nanowires started to decompose at approximately 200°C, and the superhydrophobic performance of the waterproof fire-resistant paper could be maintained below this temperature. The waterproof fire-resistant paper became hydrophilic after heating at 250°C for 1 hour, and the conversion of superhydrophobicity to superhydrophilicity occurred after heating at 300°C for 1 hour. Considering the actual working environment in daily life, the as-prepared waterproof fire-resistant paper is promising for various applications under high-temperature environments.

In addition, the superhydrophobicity of the waterproof fire-resistant paper exhibited excellent resistance to mechanical damage. Common physical damages in daily life may damage the paper surface severely and compromise the superhydrophobic performance of the paper. Various mechanical damage tests similar to the common physical damages in daily life, including the finger wipe, tape peeling, knife cutting, and sandpaper abrasion, were conducted to investigate the mechanical durability of the waterproof fire-resistant paper. Despite severe damage of the paper surface and exposure of underlying layers, the superhydrophobicity of the waterproof fire-resistant paper could be well preserved, indicating its highly stable superhydrophobicity. This was attributed to the layered hierarchical structure of the waterproof fire-resistant paper.

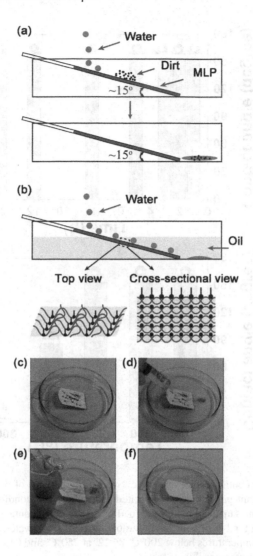

FIGURE 4.6 Self-cleaning tests for the waterproof fire-resistant paper. (a, b) Schematic illustration of the self-cleaning tests in different media: (a) in air and (b) in oil; (c) the dirt was placed on the waterproof fire-resistant paper surface before self-cleaning test; (d) the dirt was removed by water droplets from a syringe; (e) the dirt was washed away by a water flow; (f) the waterproof fire-resistant paper was clean again after the self-cleaning test. (Reprinted with permission from reference [59])

The waterproof fire-resistant paper possessed excellent resistance to high temperatures and excellent heat insulation performance, and could be used as a fire-shielding protector to protect flammable objects. In a control experiment, commercial printing paper consisting of flammable cellulose fibers was burned to ashes in 5 seconds without the protection of the waterproof fire-resistant paper, as shown in Figure 4.8a.

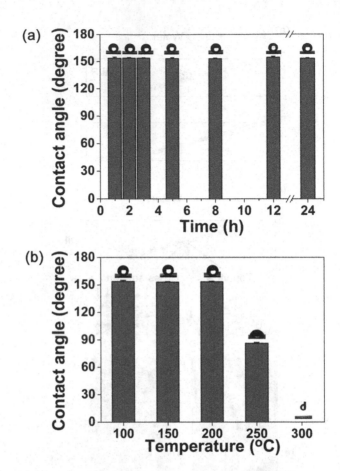

FIGURE 4.7 Water contact angles and the corresponding images of water droplets on the waterproof fire-resistant paper thermally treated under different conditions: (a) heating at 100°C for different times up to 24 hours, and in all cases the water contact angles were higher than 150°; (b) heating for 1 hour at different temperatures, and the water contact angles were higher than 150° at temperatures below 200°C, 86.22° at 250°C, and 0° at 300°C. (Reprinted with permission from reference [59])

In contrast, commercial printing paper using the waterproof fire-resistant paper as a protective layer had no obvious change after heating for 5 minutes (Figure 4.8b). Therefore, the waterproof fire-resistant paper can be used for the protection of flammable objects and substances from fire, such as important books, paper documents, and archives.

The waterproof fire-resistant paper can also be used as writing and printing paper. The experiments showed that the waterproof fire-resistant paper exhibited excellent performance in writing and printing without compromising its superhydrophobic and fire-resistant properties. The characters on the waterproof fire-resistant paper were clearly visible after heating on the flame of a spirit lamp.[59]

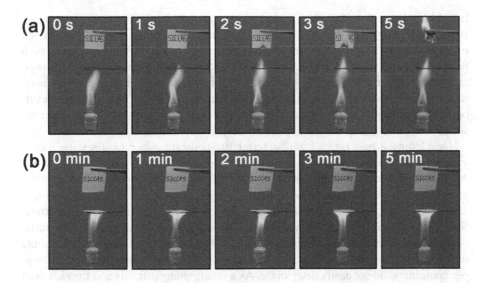

FIGURE 4.8 Fire-resistance tests of commercial printing paper without (a) and with (b) the waterproof fire-resistant paper as the protective layer. (Reprinted with permission from reference [59])

As discussed above, a new kind of highly flexible, layer-structured, superhydrophobic fire-resistant paper, made from nonflammable ultralong hydroxyapatite nanowires after surface modification with sodium oleate, has been developed. The layered structure of the waterproof fire-resistant paper enables high stability of the superhydrophobicity and excellent resistance to various mechanical damages, leading to excellent mechanical durability against physical damages. Furthermore, combining excellent fire resistance, superhydrophobicity, liquid repellency, self-cleaning property, and high thermal stability, the as-prepared waterproof fire-resistant paper is promising for various applications as recyclable adsorbent for oil-water separation, fire-shielding protector, and writing/printing paper.

4.3. ANTIBACTERIAL FIRE-RESISTANT PAPER

As we all know, the main component of traditional paper is plant fibers, which can easily absorb secretions from the human body and become a source of nutrition and cause the transmission of bacteria. In some paper sheets that are frequently used and exchanged, such as medical record paper, banknote paper, and various bills, bacteria are easily attached to the surface of the paper and spread through the crowd, causing serious problems to human health. Some paper products used in the medical and health fields require the antibacterial function in addition to sterilization. If food packaging paper is antibacterial, it can extend the preservation time of the packaged foods, fruits, and vegetables, and retard deterioration. Some types of household paper, such as napkins, makeup remover paper, and toilet paper, are required to be safe for human skin. However, paper without the antibacterial function is prone to breed bacteria in humid environments, and its use will have adverse consequences. In addition, many important library books and cultural relics are gradually destroyed by mold during the long-term preservation process. If paper itself is resistant to both bacteria and fire, precious books, documents, archives, calligraphy works, paintings, and other paper materials can be stored safely for a long time, and people do not need to worry about them getting damaged by fire, bacteria, or mold. Therefore, the development of a new type of antibacterial fire-resistant paper has important research significance and high practical application value.

Developing a new type of antibacterial fire-resistant paper requires the use of antibacterial agents in combination with ultralong hydroxyapatite nanowires. Antibacterial agents are substances or products with antibacterial and bactericidal properties that can keep the growth or reproduction of certain microorganisms (bacteria, fungi, viruses, etc.) at a sufficiently low level for a certain period of time. Silver is well known for its high antibacterial activity for a wide range of bacteria. Antibacterial materials have been used since ancient times. In ancient times, people realized that silver had an antibacterial effect, and that water stored in silver and copper containers did not easily deteriorate. As a result, many officials and the rich used silver chopsticks while dining, and people also wore silver jewelry.

The author's research group developed a highly flexible antibacterial fire-resistant paper with high antibacterial activity made from silver nanoparticle-decorated ultralong hydroxyapatite nanowires.[60] Silver nanoparticle-decorated ultralong hydroxyapatite nanowires were synthesized by the calcium oleate precursor solvothermal method using an aqueous solution containing $Ca(NO_3)_2$, $AgNO_3$, $NaOH$, and NaH_2PO_4 in a mixture of oleic acid and ethanol at 180°C for 24 hours. The SEM and TEM images in Figure 4.9a–d show that the product consisted of ultralong hydroxyapatite nanowires with diameters of approximately 20 nm and lengths of up to several hundred micrometers. In many cases, the ultralong hydroxyapatite nanowires self-assembled into nanowire bundles. The surface of the ultralong hydroxyapatite nanowires was decorated with abundant and well-dispersed silver nanoparticles with an average size of 22.5 nm and a narrow particle size distribution (Figure 4.9e). The highly flexible antibacterial fire-resistant paper was prepared using silver nanoparticle-decorated ultralong hydroxyapatite nanowires as the raw material by vacuum-assisted filtration, and the digital images of the antibacterial fire-resistant paper are shown as insets in Figure 4.9a and c. The brown color of the antibacterial fire-resistant paper indicated the successful incorporation of silver nanoparticles in the paper. The as-prepared antibacterial fire-resistant paper exhibited high flexibility, and it could be wrapped around a glass rod (Figure 4.9f). Figure 4.10 shows that the color of the as-prepared antibacterial fire-resistant paper changed gradually from light grey to brown when the content of silver nanoparticles increased.

The silver ions (Ag^+) release behavior of the antibacterial fire-resistant paper in phosphate buffered saline was investigated. Figure 4.11 shows the Ag^+ ion release profiles of four antibacterial fire-resistant paper sheets with increasing content of silver nanoparticles from (a) to (d). All samples showed an obvious initial rapid Ag^+ ion release on the first day and then a slow sustained release over the following six days. The Ag^+ release behavior was strongly dependent on the content of

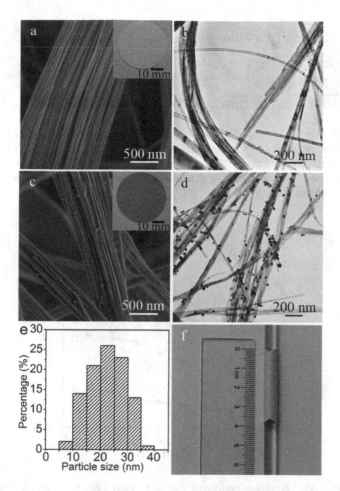

FIGURE 4.9 (a–d) SEM and TEM images: (a, b) ultralong hydroxyapatite nanowires; (c, d) silver nanoparticle-decorated ultralong hydroxyapatite nanowires (insets are the digital images of the corresponding fire-resistant paper). (e) Particle size distribution of silver nanoparticles; (f) a digital image showing that the antibacterial fire-resistant paper had a high flexibility, and could be wrapped around a glass rod. (Reprinted with permission from reference [60])

silver nanoparticles. The cumulative released Ag+ ion concentration over seven days for four samples was approximately 2.08, 4.94, 7.32, and 8.98 ppm, respectively. Furthermore, Ag+ ion release was still significant after seven days. These experimental results indicate that the as-prepared antibacterial fire-resistant paper is promising for high-performance sustained antimicrobial applications.

The antibacterial activity of the antibacterial fire-resistant paper was evaluated for both Gram-negative *Escherichia coli* (*E. coli*) and Gram-positive *Staphylococcus aureus* (*S. aureus*) by the disk diffusion method and the colony-forming unit (CFU) plate-counting method. The antibacterial fire-resistant paper with a disc shape and a

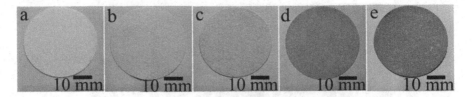

FIGURE 4.10 Digital images of the highly flexible antibacterial fire-resistant paper sheets with different contents of silver nanoparticles. (a) Fire-resistant paper made from ultralong hydroxyapatite nanowires without silver nanoparticles (white color); (b–e) antibacterial fire-resistant paper sheets with increasing content of silver nanoparticles from (b) to (e). (Reprinted with permission from reference [60])

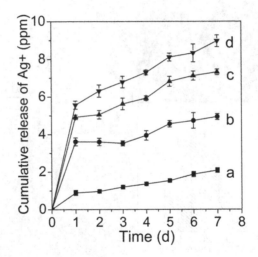

FIGURE 4.11 Ag$^+$ ion release profiles of four antibacterial fire-resistant paper sheets with increasing content of silver nanoparticles from (a) to (d). (Reprinted with permission from reference [60])

diameter of ~10 mm was placed on a confluent culture of *E. coli* and *S. aureus* on an agar plate and incubated for 24 hours. As shown in Figure 4.12, there was no inhibition zone around the fire-resistant paper sheets made from ultralong hydroxyapatite nanowires without silver nanoparticles or with a low content of silver nanoparticles, whereas an obvious inhibition zone was observed around the antibacterial fire-resistant paper with a higher content of silver nanoparticles, and the diameter of the inhibition zone increased with increasing content of silver nanoparticles.[60]

The antibacterial activity of the antibacterial fire-resistant paper was quantitatively evaluated by measuring the antibacterial efficiency. The bacteria were cultured with the antibacterial fire-resistant paper for different times and re-cultured on the agar plate and measured by the plate-counting method. As shown in Figure 4.13a and b, for *E. coli* and *S. aureus* the antibacterial activity of the fire-resistant paper

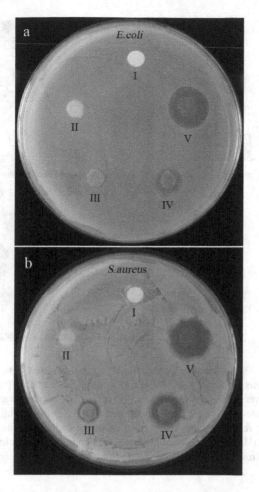

FIGURE 4.12 Digital images of the inhibition zones of the antibacterial fire-resistant paper sheets against *E. coli* (a) and *S. aureus* (b). (I) Fire-resistant paper made from ultralong hydroxyapatite nanowires without silver nanoparticles; (II–V) antibacterial fire-resistant paper sheets with increasing content of silver nanoparticles from (II) to (V). (Reprinted with permission from reference [60])

made from ultralong hydroxyapatite nanowires without silver nanoparticles was less than 10%. However, the antibacterial activity of the antibacterial fire-resistant paper increased significantly with increasing content of silver nanoparticles and was time-dependent. All *E. coli* were killed by the antibacterial fire-resistant paper sheets with over 2 wt.% silver loading within 12 hours. Similarly, more than 99% of *S. aureus* were killed by the antibacterial fire-resistant paper with a silver content of 2.25 wt.% within 24 hours (Figure 4.13c). The experiments suggested that the antibacterial fire-resistant paper exhibited a better antimicrobial effect against Gram-negative *E. coli* than Gram-positive *S. aureus*.[60]

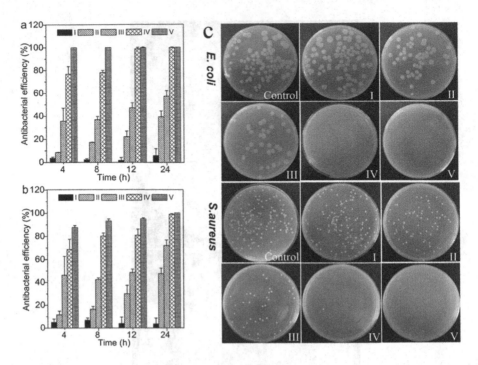

FIGURE 4.13 (a, b) Antibacterial efficiencies for *Escherichia coli* (*E. coli*) (a) and *Staphylococcus aureus* (*S. aureus*) (b) by the antibacterial fire-resistant paper sheets with different contents of silver nanoparticles as a function of time; (c) digital images of re-cultured *E. coli* and *S. aureus* colonies on agar plates incubated with the antibacterial fire-resistant paper for 24 hours. (I) fire-resistant paper made from ultralong hydroxyapatite nanowires without silver nanoparticles; (II–V) antibacterial fire-resistant paper sheets with increasing content of silver nanoparticles from (II) to (V). (Reprinted with permission from reference [60])

In another related research work, an ultralong hydroxyapatite nanowires-based dual antibacterial paper co-loaded with silver nanoparticles and antibiotic was prepared.[56] Ultralong hydroxyapatite nanowires were used to prepare silver nanoparticles on the surface *in situ* using an aqueous solution containing AgNO$_3$ under sunlight in the absence of any additional reducing agent at room temperature. Subsequently, ciprofloxacin, as an antibiotic, was loaded on the ultralong hydroxyapatite nanowires loaded with silver nanoparticles, as shown in Figure 4.14a. In contrast to the white fire-resistant paper consisting of only ultralong hydroxyapatite nanowires without any antibacterial agent, the dual antibacterial fire-resistant paper consisting of ultralong hydroxyapatite nanowires co-loaded with silver nanoparticles and ciprofloxacin had a brown color (Figure 4.14b). The dual antibacterial paper exhibited a high flexibility and could maintain its shape without visible damage after repeated bending to nearly 180° or rolling around a glass rod (Figure 4.14c and d). The dual antibacterial fire-resistant paper exhibited unique properties, including high specific surface area, high drug loading capacity, good biocompatibility, sustained and

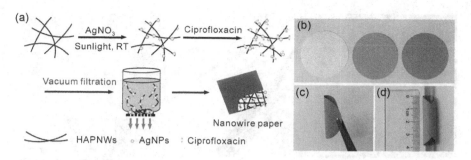

FIGURE 4.14 (a) Schematic illustration of the preparation process of the dual antibacterial fire-resistant paper consisting of ultralong hydroxyapatite nanowires co-loaded with silver nanoparticles and ciprofloxacin; (b) digital images of the corresponding fire-resistant paper sheets with a diameter of 4.0 cm consisting of only ultralong hydroxyapatite nanowires without any antibacterial agent, or ultralong hydroxyapatite nanowires loaded with silver nanoparticles, or ultralong hydroxyapatite nanowires co-loaded with silver nanoparticles and ciprofloxacin (from left to right); (c, d) the dual antibacterial fire-resistant paper consisting of ultralong hydroxyapatite nanowires co-loaded with silver nanoparticles and ciprofloxacin exhibited a high flexibility. (Reprinted with permission from reference [56])

pH-responsive drug release behavior, long-time high antibacterial activity, and good recycling performance.

Figure 4.15 shows SEM and TEM images of the as-prepared ultralong hydroxyapatite nanowires and fire-resistant paper sheets. One can see that silver nanoparticles with an average size of about 50 nm were well-dispersed and anchored on the surface of ultralong hydroxyapatite nanowires. SEM observation was also carried out to investigate the morphology of the cross-section of the fire-resistant paper. As shown in Figure 4.15c and f, both the fire-resistant paper composed of only ultralong hydroxyapatite nanowires and the antibacterial fire-resistant paper consisting of ultralong hydroxyapatite nanowires loaded with silver nanoparticles showed a well-defined layered structure. The layered structure was favorable for the storage of ciprofloxacin with a high drug loading capacity. After loading with ciprofloxacin, the paper could still maintain a layered structure but with a reduced interlayer spacing (Figure 4.15i). This phenomenon may be explained by the interactions between ciprofloxacin molecules and ultralong hydroxyapatite nanowires in two ways: chelating carboxyl groups in ciprofloxacin molecules with Ca^{2+} ions in ultralong hydroxyapatite nanowires, hydrogen bonding between the carboxyl groups/imino groups in ciprofloxacin molecules and the hydroxylated surface of ultralong hydroxyapatite nanowires. As a result, these interactions reduced the spacing between the layers of the paper.

The release behaviors of Ag^+ ions and ciprofloxacin in phosphate buffered saline solutions with different pH values were studied, and the experimental results are shown in Figure 4.16. Figure 4.16a shows the release profiles of ciprofloxacin from the as-prepared dual antibacterial fire-resistant paper in phosphate buffered saline solutions with different pH values. Ciprofloxacin release was relatively rapid at the early stage at three different pH values. At the physiological pH value (i.e., 7.4),

FIGURE 4.15   SEM and TEM images of the as-prepared ultralong hydroxyapatite nanowires and the cross section of fire-resistant paper sheets. (a–c) Ultralong hydroxyapatite nanowires without silver nanoparticles, and a fire-resistant paper with a layered structure; (d–f) ultralong hydroxyapatite nanowires with silver nanoparticles, and an antibacterial fire-resistant paper with a layered structure; (g–i) ultralong hydroxyapatite nanowires co-loaded with silver nanoparticles and ciprofloxacin, and a dual antibacterial fire-resistant paper with a layered structure. (Reprinted with permission from reference [56])

ciprofloxacin was released slowly, and sustained drug release performance was observed. However, at a lower pH value, ciprofloxacin was released faster. At day 8, about 76.4%, 70.1%, and 37.7% of ciprofloxacin was released at the pH values of 4.5, 6.0, and 7.4, respectively. The ciprofloxacin release profile showed a pH-responsive feature. The accelerated drug release under acidic conditions may be due to weaker interactions between ciprofloxacin molecules and ultralong hydroxyapatite nanowires and the partial dissolution of the drug carrier.

An accelerated effect was found on further study of the release profiles of Ag^+ ions. Figure 4.16b shows the release profiles of Ag^+ ions from the dual antibacterial

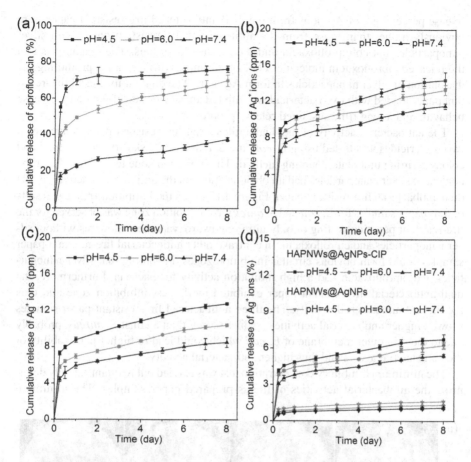

FIGURE 4.16 *In vitro* release profiles in phosphate buffered saline solutions with different pH values. (a) Cumulative release percentage of ciprofloxacin from the dual antibacterial fire-resistant paper; (b, c) concentrations of cumulative released Ag^+ ions from the dual antibacterial fire-resistant paper (b), and the antibacterial fire-resistant paper with silver nanoparticles but no ciprofloxacin (c); (d) cumulative percentage of released Ag^+ ions from the dual antibacterial fire-resistant paper, and the antibacterial paper with silver nanoparticles but no ciprofloxacin. (Reprinted with permission from reference [56])

paper at different pH values. The release of Ag^+ ions was rapid in the first 12 hours but slow and sustained in the following seven days. Furthermore, the Ag^+ ions release profile showed a pH-responsive feature, and higher concentrations of released Ag^+ ions could be obtained at lower pH values. In order to investigate the accelerated effect of ciprofloxacin on Ag^+ ion release, the release profiles of Ag^+ ions from the antibacterial fire-resistant paper with silver nanoparticles but no ciprofloxacin were studied (Figure 4.16c), and it was found that the Ag^+ ion release profile also exhibited a pH-responsive behavior and that the cumulative released concentrations of Ag^+ ions in eight days at three different pH values were lower than those of the dual antibacterial fire-resistant paper. Furthermore, Figure 4.16d shows that the cumulative release percentages of Ag⁺ ions from the dual antibacterial fire-resistant paper were obviously higher than those from the antibacterial fire-resistant paper with silver nanoparticles but no ciprofloxacin under the same conditions. The reason was that the released ciprofloxacin molecules could coordinate with Ag^+ ions, promoting the dissolution of silver nanoparticles to form Ag^+ ions. The dual antibacterial fire-resistant paper loaded with two bactericides exhibited an accelerated effect on the release behavior and a long-time sustained release profile.[56]

The antibacterial activity of the dual antibacterial fire-resistant paper loaded with two bactericides was tested using *E. coli* and *S. aureus* by disk diffusion method and colony-forming unit plate-counting method. The inhibition zone tests showed that the content of silver nanoparticles had an obvious effect on the antibacterial activity of the dual antibacterial fire-resistant paper. Figure 4.17 shows the inhibition zone test results of four paper samples for *E. coli* and *S. aureus*. No inhibition zone was observed for the fire-resistant paper consisting of only ultralong hydroxyapatite nanowires without silver nanoparticles and ciprofloxacin. In contrast, other antibacterial fire-resistant paper samples could form obvious circular inhibition zones, indicating that silver nanoparticles and ciprofloxacin had a high inhibition activity for bacteria. Furthermore, the dual antibacterial fire-resistant paper exhibited the largest inhibition zone and thus the highest antibacterial activity. The three antibacterial fire-resistant paper samples showed higher antibacterial activities against *E. coli* than against *S. aureus*, probably due to the thin outer membrane of *E. coli* which could lead to higher permeability of the antibacterial agents and thus higher antibacterial activity.

The minimum inhibitory concentration test was carried out to quantitatively determine the antibacterial activities of the as-prepared paper samples. The minimum

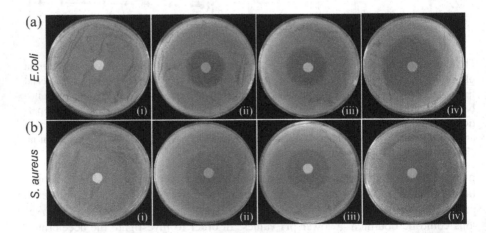

FIGURE 4.17 Digital images of the inhibition zones of four paper samples against: (a) *E. coli*, and (b) *S. aureus*. (i) Ultralong hydroxyapatite nanowires; (ii) the antibacterial fire-resistant paper with silver nanoparticles but no ciprofloxacin; (iii) the antibacterial fire-resistant paper with ciprofloxacin but no silver nanoparticles; (iv) the dual antibacterial fire-resistant paper with both silver nanoparticles and ciprofloxacin. (Reprinted with permission from reference [56])

inhibitory concentration values of three antibacterial fire-resistant paper sheets (consisting of silver nanoparticles, ciprofloxacin, and silver nanoparticles + ciprofloxacin, respectively) were 60, 75, 30 μg mL^{-1} against *E. coli*; and 75, 100, 40 μg mL^{-1} against *S. aureus*, respectively. The normalized minimum inhibitory concentration values of the antibacterial fire-resistant paper consisting of ultralong hydroxyapatite nanowires and silver nanoparticles against *E. coli* and *S. aureus* were 3.05 and 3.81 μg mL^{-1}, respectively, which were lower than those of only silver nanoparticles, showing that the antibacterial fire-resistant paper consisting of ultralong hydroxyapatite nanowires and silver nanoparticles had a higher antibacterial activity than that consisting of only silver nanoparticles at the same silver content. The dual antibacterial fire-resistant paper had the lowest minimum inhibitory concentration value, indicating the highest antibacterial activity among three antibacterial fire-resistant paper samples.[56]

The consecutive recycling performance and long-time stability of the dual antibacterial fire-resistant paper with both silver nanoparticles and ciprofloxacin were also investigated. As shown in Figure 4.18a, after eight cycles, the dual antibacterial fire-resistant paper could maintain more than 95% of its original antibacterial activity against both *E. coli* and *S. aureus*. As discussed above, the dual antibacterial fire-resistant paper exhibited long-time and sustained release of Ag$^+$ ions and ciprofloxacin, and can be used as a reusable antibacterial fire-resistant paper with synergistic high antibacterial activity.

The long-time stability testing results of the dual antibacterial fire-resistant paper are shown in Figure 4.18b; the dual antibacterial fire-resistant paper could preserve more than 95% of its original antibacterial activity for both *E. coli* and *S. aureus*, showing the excellent long-time antibacterial stability. In addition, the dual antibacterial fire-resistant paper was incubated with a liquid broth and continuously stained by bacteria every 24 hours, and the experiment indicated that the dual antibacterial fire-resistant paper could preserve 100% antibacterial efficiency after five continuous cycles for both *E. coli* and *S. aureus* (Figure 4.18c and d).[56]

4.4. WATERPROOF ELECTRICALLY CONDUCTIVE FIRE-RESISTANT PAPER

In today's modern society, electronic devices are everywhere, ranging from mobile phones to televisions to precision instruments and equipment. Electronic devices have greatly changed people's work and lifestyle. Owing to their excellent performance and functions, ability to adapt to physical deformation, and ease of use, flexible electronic devices and wearable electronic devices have become popular in recent years, and the demands for lightweight, portable, and flexible electronic devices are ever-increasing. However, traditional hard and brittle electronic components based on silicon and indium tin oxide are difficult to meet the requirements of high flexibility and the ability to work in harsh environments. The performance of common electronic devices can be easily affected by the surrounding environment. For example, metals are the first choice for conductive materials, but many metals are easily oxidized and corroded, which may degrade their properties and even result in failure

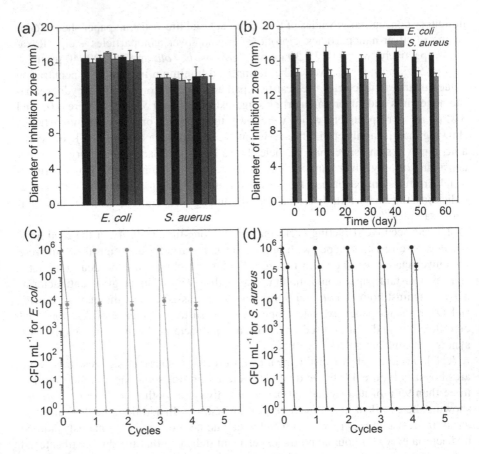

FIGURE 4.18 (a) Consecutive recycling testing results of the dual antibacterial fire-resistant paper with both silver nanoparticles and ciprofloxacin against *E. coli* and *S. aureus* for different cycles by the inhibition zone method; (b) long-time stability testing results of the dual antibacterial fire-resistant paper with both silver nanoparticles and ciprofloxacin against *E. coli* and *S. aureus* for two months; (c, d) five cycles of antibacterial tests of the dual antibacterial fire-resistant paper with both silver nanoparticles and ciprofloxacin against *E. coli* (c) and *S. aureus* (d) by the liquid broth culture method. At each initial point, fresh bacterial fluid was added to the liquid broth and each cycle time was 24 hours. (Reprinted with permission from reference [56])

in functions over time. Furthermore, contaminants, moisture, freezing, and snowing can deteriorate the performance of electronic devices. Electronic devices are also easily destroyed by floods, high temperatures, or fire. Icing and snowing are common but extremely harsh environments. For example, a serious snow storm was witnessed in South China in 2008; the resulting excessive ice accretion on the electrical transmission lines and power network towers caused severe damages and enormous economic losses. Fortunately, a superhydrophobic surface can inhibit the accumulation of water and ice. In addition, a combination of superhydrophobic surface and electrothermal effect can enhance the anti-icing efficiency. One severe problem is that the transmission of electrical power may cause fire accidents due to short circuit, overload, poor contact, overheating, electric leakage, misuse, etc. The fire disasters can inevitably damage or even destroy the electronic devices containing flammable materials. Therefore, fire resistance is highly desirable in flexile electronic devices so that they can withstand high temperatures and fire.

Although inorganic ultralong hydroxyapatite nanowires are electrically insulating, their high flexibility and nonflammability are attractive for use in fire-resistant flexible electronic devices. As discussed above, the waterproof fire-resistant paper was developed using ultralong hydroxyapatite nanowires modified with sodium oleate.[59] Unfortunately, the decomposition of oleate at high temperatures (>200°C) is not suitable for electrothermal applications.

Recently, the author's research group developed a new kind of flexible waterproof electrically conductive fire-resistant paper with electrothermal property based on ultralong hydroxyapatite nanowires, Ketjen black particles, and poly(dimethylsiloxane).[61] Ketjen black particles were used as the electrically conductive material, and poly(dimethylsiloxane) was used as a superhydrophobic coating on the paper surface. The superhydrophobic surface could enable the as-prepared electrically conductive fire-resistant paper to be resistant to corrosion (pH 1–14) and moisture (up to 90%), and the superhydrophobic surface showed a high thermal stability (up to 300°C). The waterproof property could enable the electrically conductive fire-resistant paper to work steadily even when it was immersed in water. In addition, the synergistic effect between the superhydrophobic and electrothermal properties could promote the removal of water and de-icing on electrically conductive fire-resistant paper. More importantly, the as-prepared waterproof electrically conductive fire-resistant paper showed excellent flame-retardant performance, and it could work stably and continuously in a flame.

As shown in Figure 4.19a, the as-prepared electrically conductive fire-resistant paper without a poly(dimethylsiloxane) coating was superhydrophilic, with a water contact angle of ~0°. The coating of poly(dimethylsiloxane) on the electrically conductive fire-resistant paper had no obvious effect on its appearance, because poly(dimethylsiloxane) was transparent, but could obviously change the surface wettability, making it superhydrophobic, with a water contact angle of >150° (Figure 4.19d). However, the paper consisting of ultralong hydroxyapatite nanowires with a poly(dimethylsiloxane) coating without the addition of Ketjen black particles exhibited hydrophobicity, with a lower water contact angle (137.3 ± 0.6). Because surface wettability is determined by both surface energy and microstructure, Ketjen black particles played an important role in regulating the surface wettability of the paper, and the addition of Ketjen black particles could form a beads-on-string structure and increase the roughness (Figure 4.19b). The increased roughness was enough to offset the negative impact of poly(dimethylsiloxane) (Figure 4.19e), leading to the superhydrophobicity. In addition, compared with the electrically conductive fire-resistant paper consisting of 20 wt.% Ketjen black particles and ultralong hydroxyapatite nanowires without a poly(dimethylsiloxane) coating, an obvious Si peak was detected in the electrically conductive fire-resistant paper after poly(dimethylsiloxane) coating by energy-dispersive X-ray spectroscopy (Figure 4.19c and f). As shown in Figure 4.19g,

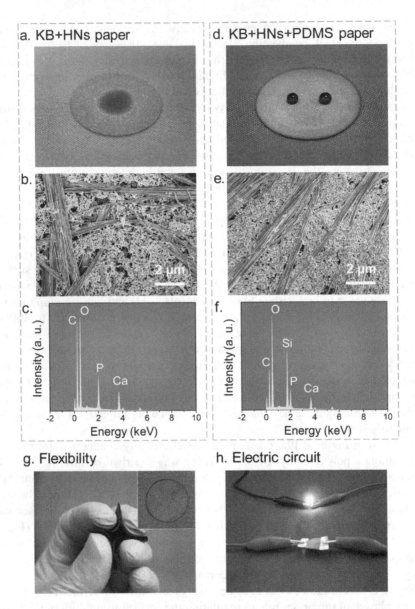

FIGURE 4.19 (a) Digital image, (b) SEM image, and (c) energy-dispersive X-ray spectroscopy of the waterproof electrically conductive fire-resistant paper containing 20 wt.% Ketjen black (KB) particles and 80 wt.% ultralong hydroxyapatite nanowires (HNs); (d) digital image, (e) SEM image, and (f) energy-dispersive X-ray spectroscopy of the waterproof electrically conductive fire-resistant paper containing 20 wt.% Ketjen black particles and 80 wt.% ultralong hydroxyapatite nanowires with a poly(dimethylsiloxane) coating; (g) the waterproof electrically conductive fire-resistant paper could be twisted without breaking (the inset shows the paper after twisting), showing a high flexibility; (h) the waterproof electrically conductive fire-resistant paper was connected with a direct current power supply at a voltage of 3 V and an light-emitting diode lamp, which was continuously lighting. (Reprinted with permission from reference [61])

the waterproof electrically conductive fire-resistant paper was highly flexible and could be twisted and bent without breaking.

The electrical properties of the waterproof electrically conductive fire-resistant paper sheets with different contents of Ketjen black particles were investigated. In the experiments, the paper was connected with a light-emitting diode (LED) lamp at an applied voltage of 3 V (Figure 4.19h). The electrically insulated fire-resistant paper without Ketjen black particles could not light the LED lamp; however, the LED lamp could continuously light up in the presence of the waterproof electrically conductive fire-resistant paper. The experimental results indicated that the electrical conductivity of the waterproof electrically conductive fire-resistant paper increased with increasing content of Ketjen black particles.

The mechanical properties of the waterproof electrically conductive fire-resistant paper were investigated. The addition of Ketjen black particles could enhance the tensile strength of ultralong hydroxyapatite nanowires. For example, the waterproof electrically conductive fire-resistant paper (10 wt.% Ketjen black particles) had a higher tensile strength (2.82 ± 0.50 MPa) than the paper without Ketjen black particles (1.71 ± 0.99 MPa). However, the tensile strength of the waterproof electrically conductive fire-resistant paper decreased by further increasing the content of Ketjen black particles. The addition of Ketjen black particles also resulted in decreased strain at failure and higher Young's modulus of the waterproof electrically conductive fire-resistant paper. The waterproof electrically conductive fire-resistant paper (30 wt.% Ketjen black particles) had the highest Young's modulus (141.33 ± 11.68 MPa).

The electrical performance of the waterproof electrically conductive fire-resistant paper was tested under bending cycles at a bending angle of $\sim 90°$. The experimental results showed that the electrical performance was relatively stable during 500 bending cycles, and the waterproof electrically conductive fire-resistant paper was not broken after 500 bending cycles, indicating its high flexibility and good electrical and mechanical properties.

The electrical performance of the waterproof electrically conductive fire-resistant paper underwater was also evaluated, as shown in Figure 4.20. It was difficult to immerse the waterproof electrically conductive fire-resistant paper into water because of its superhydrophobicity, and an external force was needed. The electrical current values were real-time recorded after the waterproof electrically conductive fire-resistant paper was immersed in water, and the results are shown in Figure 4.20a and b. The testing process in the first 10 seconds is also shown in Figure 4.20c. Although the electrical current increased gradually, very little change of electrical current was observed during the testing process, and the relative electrical current change was 3.65% during the whole process. The high stability of the electrical current and the structural integrity enabled the waterproof electrically conductive fire-resistant paper to work steadily underwater (Figure 4.20d). No obvious change in the brightness of the LED lamp was observed during the testing time period, indicating a stable and continuous underwater service of the waterproof electrically conductive fire-resistant paper.

The electrothermal property of the waterproof electrically conductive fire-resistant paper was also investigated. The experiments indicated that both the content of

a. Setup

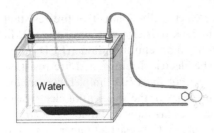

b. Electrical current change

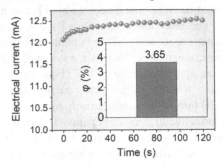

c. Real-time monitoring

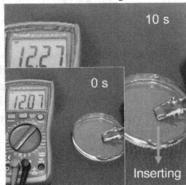

d. Underwater service

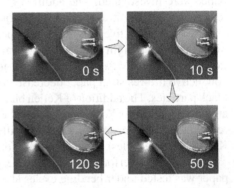

FIGURE 4.20 The underwater electrical performance of the waterproof electrically conductive fire-resistant paper. (a) Schematic illustration of the experimental setup for the waterproof electrically conductive fire-resistant paper (10 mm × 20 mm) connected with an ammeter and a direct current power supply at a voltage of 3 V (or an LED lamp); (b) electrical current through the waterproof electrically conductive fire-resistant paper after it was immersed in water, and the inset shows the relative electrical current change (φ); (c) digital images of real-time monitoring of the electrical current through the waterproof electrically conductive fire-resistant paper in water; (d) digital images of real-time monitoring of the brightness of an LED lamp connected with the underwater waterproof electrically conductive fire-resistant paper and a direct current power supply at a voltage of 3 V. (Reprinted with permission from reference [61])

Ketjen black particles and the applied voltage had effects on the electrothermal efficiency. The electrically insulated fire-resistant paper without the addition of Ketjen black particles had no thermal effect, and the temperature of the paper surface had no obvious change over time. In contrast, the waterproof electrically conductive fire-resistant paper showed the electrothermal effect. The surface temperature of the waterproof electrically conductive fire-resistant paper increased rapidly and reached the quasi-steady-state temperature quickly within 10 seconds, and then the paper surface temperature leveled off. The experimental results indicated that the higher the content of Ketjen black particles or the higher the applied voltage, the higher the paper surface temperature. The surface temperature of the waterproof electrically conductive fire-resistant paper with 20 wt.% Ketjen black particles under an applied voltage of 20 V could reach a high temperature of 224.25°C in 9.6 seconds. Such high temperature produced within a short period of time is suitable for highly efficient de-icing and water evaporation. The waterproof electrically conductive fire-resistant paper exhibited a steady electrothermal effect and good recyclability during the testing process of repeating heating/cooling cycles for many times.

The two problems related to a superhydrophobic surface were that extremely tiny water droplets adhered to the surface and ice accretion formed on the surface. The electrothermal effect of the waterproof electrically conductive fire-resistant paper can solve these two problems. In an experiment, a water droplet with a volume of 3 µL was separately deposited on the unheated waterproof electrically conductive fire-resistant paper, and on the heated one with an applied voltage of 20 V, as shown in Figure 4.21a and b. The experimental results indicated that the higher surface temperature of the heated waterproof electrically conductive fire-resistant paper could obviously reduce the evaporation time of a water droplet (128 seconds), whereas there was no obvious change in the water volume on the unheated paper even after 10 minutes. The electrothermal effect of the waterproof electrically conductive fire-resistant paper on the melting of ice is shown in Figure 4.21c–e. It took about 103 seconds for the melting and complete removal of ice on the unheated waterproof electrically conductive fire-resistant paper (Figure 4.21c). For the heated waterproof electrically conductive fire-resistant paper, the de-icing time was much shorter (23 seconds) and the de-icing efficiency was ~4.5 times that of the unheated paper (Figure 4.21d). In addition, the surface wettability also played an important role in the de-icing process. As shown in Figure 4.21e, although the heated electrically conductive fire-resistant paper without a poly(dimethylsiloxane) coating could accelerate the melting of ice, the melted water was absorbed by the paper because of its superhydrophilic property, and it took 240 seconds (~10.4 times that of the heated waterproof electrically conductive fire-resistant paper) for the heated superhydrophilic electrically conductive fire-resistant paper without a poly(dimethylsiloxane) coating to become entirely dry. The experimental results showed that the electrothermal effect could accelerate the melting of ice, and the superhydrophobic surface could enable the rapid removal of the melted water. The high temperature reached in a short time period (e.g., 224.25°C) did not degrade the superhydrophobicity of the waterproof electrically conductive fire-resistant paper owing to the good thermal stability of the poly(dimethylsiloxane) coating up to 300°C.[61]

The flame-retardant property of the waterproof electrically conductive fire-resistant paper was also investigated. The waterproof electrically conductive fire-resistant paper exhibited a high thermal stability and excellent fire-retardant performance. When the waterproof electrically conductive fire-resistant paper was exposed to fire, only a small amount of gas was observed, there was no obvious flame on the paper, and the paper could well preserve its structural integrity. Figure 4.22a shows the electrical current change of the waterproof electrically conductive fire-resistant paper in a flame. During the heating process, the electrical current increased from the initial value of 11.92–13.39 mA in 1 minute and then became stable (13.34 mA

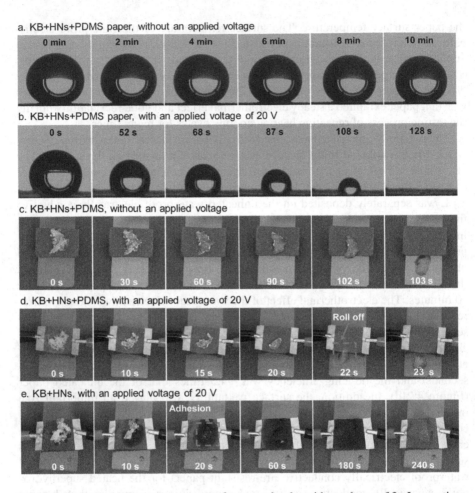

FIGURE 4.21 (a, b) Evolution process of a water droplet with a volume of 3 µL over time on the waterproof electrically conductive fire-resistant paper (KB + HNs + PDMS) without and with an applied voltage of 20 V, respectively; (c, d) evolution process of ice over time on the waterproof electrically conductive fire-resistant paper without and with an applied voltage of 20 V, respectively; (e) evolution process of ice over time on the superhydrophilic electrically conductive fire-resistant paper (20 wt.% Ketjen black particles) without poly(dimethylsiloxane) coating with an applied voltage of 20 V. (Reprinted with permission from reference [61])

at 7 minutes). The maximal current change during the heating process was 15.5% (Figure 4.22b). After the heating process, the electrical current was 10.80 mA, and the total current change was relatively low (9.40%) and the electrical current after the heating was 90.60% of the initial value (Figure 4.22b). Real-time monitoring of the electrical current of the waterproof electrically conductive fire-resistant paper during the heating process is shown in Figure 4.22c, and the real-time monitoring of the brightness of an LED lamp during the heating process for 7 minutes is shown

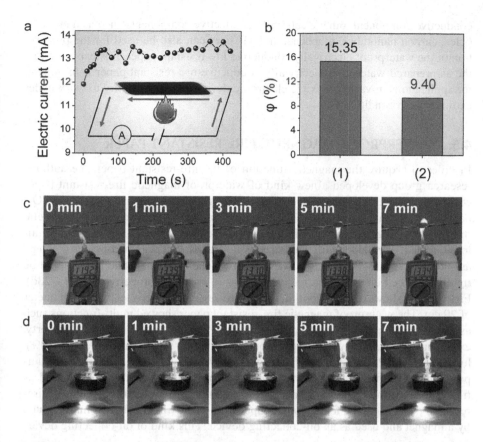

FIGURE 4.22 Real-time monitoring of electrical current through the waterproof electrically conductive fire-resistant paper exposed to the flame and connected with a direct current power supply at a voltage of 3 V and an ammeter. (a) Electrical current change during the exposure of the paper to the flame for 7 minutes; (b) relative electrical current change (φ): (1) maximal relative electrical current change and (2) total relative electrical current change; (c) digital images of the real-time monitoring of electrical current through the paper in the flame; (d) digital images of the real-time monitoring of brightness of an LED lamp connected with a direct current power supply at a voltage of 3 V and the paper in the flame. (Reprinted with permission from reference [61])

in Figure 4.22d. The experimental results indicated that the waterproof electrically conductive fire-resistant paper could work normally and continuously under extreme conditions of heating in a flame, and the brightness of the LED lamp was as high as the original state during the whole heating process.

As discussed above, the flexible waterproof electrically conductive fire-resistant paper possessed excellent superhydrophobicity, good electrical conductivity, high thermal stability, and excellent flame retardancy and electrothermal effect, and it could work well under extreme conditions, including underwater and in the flame. It should be pointed out that in addition to Ketjen black particles as the electrically conductive component, other electrically conductive components such as metal particles, carbon nanotubes, graphene, and MXenes can also be used for the preparation of the waterproof electrically conductive fire-resistant paper. It is expected that the as-prepared waterproof electrically conductive fire-resistant paper is promising for applications in various flexible electronic devices with the ability to work under extreme harsh conditions.[61]

4.5. WATERPROOF MAGNETIC FIRE-RESISTANT PAPER

In order to acquire the magnetic function of the fire-resistant paper, the author's research group developed a new kind of waterproof magnetic fire-resistant paper consisting of ultralong hydroxyapatite nanowires, magnetic iron oxide (Fe_3O_4) nanoparticles, and a poly(dimethylsiloxane) coating.[62] The waterproof magnetic fire-resistant paper had high flexibility and excellent processability, and could be tailored, folded, or rolled readily into various desired shapes without obvious damage, as shown in Figure 4.23a–c. The waterproof magnetic fire-resistant paper could be tightly absorbed on a magnet, indicating a good magnetic property (Figure 4.23d). Figure 4.23e shows a waterproof magnetic fire-resistant paper sheet with a diameter of 20 cm. The waterproof magnetic fire-resistant paper showed multi-functions such as excellent superhydrophobicity, high thermal stability, good magnetic property, excellent fire resistance, high separation performance for oil and water (>99.0%), high water flux (2,924.3 L m^{-2} h^{-1}), and good recycling ability. In addition, the waterproof magnetic fire-resistant paper had a porous structure, and could be used as filter paper for the oil-water separation. In addition, the waterproof magnetic fire-resistant paper could be folded into a paper mini-boat, which could be manipulated by a magnet and used as an oil-collecting device. This kind of oil-collecting device

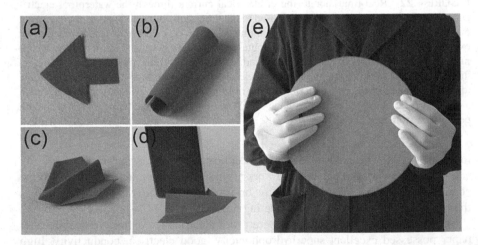

FIGURE 4.23 (a–e) Digital images of the as-prepared waterproof magnetic fire-resistant paper with high flexibility and good processability, and the paper exhibited a brown color, and could be tightly attracted by a magnet. (Reprinted with permission from reference [62])

could achieve integrated magnetic-driven continuous oil-water separation and oil collection.

The waterproof magnetic fire-resistant paper exhibited a high thermal stability and nonflammability. As shown in Figure 4.24a, the superhydrophobicity of the waterproof magnetic fire-resistant paper was highly stable after treatment in a wide temperature range from -198°C to 250°C for 1 hour. However, when the temperature

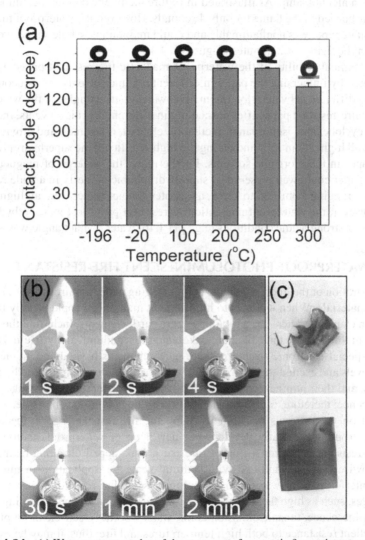

FIGURE 4.24 (a) Water contact angles of the waterproof magnetic fire-resistant paper after treatment at different temperatures for 1 hour; (b) fire-resistance tests of the common cellulose fiber paper (first row) and the waterproof magnetic fire-resistant paper (second row); (c) digital images of the common cellulose fiber paper (first row) and the waterproof magnetic fire-resistant paper (second row) after the fire-resistance test. (Reprinted with permission from reference [62])

was increased to 300°C, the water contact angle decreased to 133.24 ± 3.72°, indicating that the paper lost its superhydrophobicity and was only hydrophobic because of the decomposition of poly(dimethylsiloxane). These experimental results indicate the excellent superhydrophobic stability of the waterproof magnetic fire-resistant paper and its potential applications under harsh conditions. In testing the fire-resistant performance, the waterproof magnetic fire-resistant paper was directly heated in the flame of an alcohol lamp. As illustrated in Figure 4.24b, the common cellulose fiber paper was burned in the flame for only 4 seconds. However, the waterproof magnetic fire-resistant paper was nonflammable and could maintain its whole shape even after heating in the flame for 2 minutes (Figure 4.24c).[62]

The chemical stability of the waterproof magnetic fire-resistant paper was also investigated by immersing the paper in different organic solvents and aqueous solutions with different pH values for 1 hour. The water contact angles of the waterproof magnetic fire-resistant paper after immersing in different organic solvents, including acetone, cyclohexane, isopropanol, methanol, ethanol, n-hexanol, and ethylene glycol, were all higher than 150°, indicating the high stability of the superhydrophobicity of the paper in these organic solvents. Furthermore, the waterproof magnetic fire-resistant paper could well preserve its superhydrophobic property in a wide range of pH values, ranging from 2.35 to 12.98, and water contact angles were all higher than 150°. However, the waterproof magnetic fire-resistant paper lost its superhydrophobicity under strong acidic conditions (e.g., pH 1.46, water contact angle was ~132°).

4.6. WATERPROOF PHOTOLUMINESCENT FIRE-RESISTANT PAPER

The preparation of the fluorescent anti-counterfeiting paper requires the use of luminescent materials. When it comes to luminescent materials, people usually think of rare earth elements. There are a total of 17 rare earth elements, including the 15 lanthanides in the periodic table of chemical elements, scandium, and yttrium. Because of their special electronic structures, rare earth elements have abundant electronic energy levels and excited states, presenting spectral properties unmatched by general elements, and their luminescence wavelengths almost cover the entire range of solid luminescence; therefore, rare earth elements are a huge luminous treasure.

Based on the previous research work on the fire-resistant paper, the author's research group successfully developed a new type of waterproof photoluminescent fire-resistant paper consisting of rare earth ions–doped ultralong hydroxyapatite nanowires surface modified with sodium oleate for application in multimode anti-counterfeiting. The waterproof photoluminescent fire-resistant paper has many advantages, such as high flexibility, good processability, writing and printing ability, photoluminescence, tunable emission color, waterproofness, self-cleaning property, and excellent resistance to both high temperatures and fire; thus, it may be exploited for developing a novel multimode anti-counterfeiting technology for various high-level-security anti-forgery applications.[63]

Figure 4.25 shows the digital images of the as-prepared photoluminescent fire-resistant paper consisting of ultralong hydroxyapatite nanowires doped with 5 mol% Tb^{3+} ions. Similar to the undoped fire-resistant paper and Eu^{3+}-doped

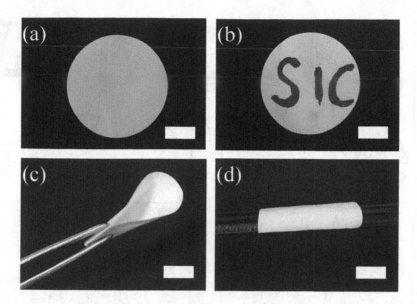

FIGURE 4.25 Digital images of the as-prepared photoluminescent fire-resistant paper consisting of 5 mol% Tb^{3+} ions-doped ultralong hydroxyapatite nanowires. (a) A circular photoluminescent fire-resistant paper; (b) the photoluminescent fire-resistant paper could be used for writing; (c) and (d) the photoluminescent fire-resistant paper exhibited a high flexibility. The scale bars are 1 cm. (Reprinted with permission from reference [63])

photoluminescent fire-resistant paper, the photoluminescent fire-resistant paper doped with Tb^{3+} ions showed a high-quality white color (Figure 4.25a). The photoluminescent fire-resistant paper was writable and printable, and text could be easily and clearly written on the paper by using a Chinese brush (Figure 4.25b). In addition, the photoluminescent fire-resistant paper exhibited a high flexibility (Figure 4.25c and d). A bending test was carried out to investigate the bending durability of the photoluminescent fire-resistant paper. The experiments indicated that the paper could be well preserved without any obvious damage after repeated bending or rolling around a glass bar. In addition, a slight crease was observed on the surface of the photoluminescent fire-resistant paper after being bent by 135° 500 times. The SEM observation showed the layered structure of the cross-section (thickness direction) of the photoluminescent fire-resistant paper. This is a common phenomenon observed on the cross-section of the fire-resistant paper based on ultralong hydroxyapatite nanowires.[56, 57, 59, 64]

The photoluminescence excitation and emission spectra of the as-prepared photoluminescent fire-resistant paper at room temperature are shown in Figure 4.26. According to the excitation spectra in Figure 4.26a and b, the most intense excitation peaks at 378 and 393 nm were chosen as the excitation wavelengths of the photoluminescent fire-resistant paper doped with Tb^{3+} and Eu^{3+} ions, respectively. Four emission peaks at 489, 545, 587, and 620 nm were observed in the photoluminescence emission spectrum of the Tb^{3+}-doped photoluminescent fire-resistant paper

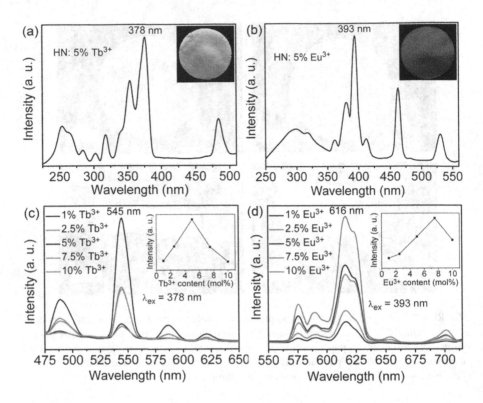

FIGURE 4.26 Photoluminescence properties of the photoluminescent fire-resistant paper. (a) Excitation spectrum of the photoluminescent fire-resistant paper doped with 5 mol% Tb^{3+} ions; (b) excitation spectrum of the photoluminescent fire-resistant paper doped with 5 mol% Eu^{3+} ions (the insets in (a) and (b) are the digital images of the corresponding paper sheets under irradiation of a UV lamp (~365 nm), emitting a green and red light, respectively); (c) photoluminescence emission spectra of the photoluminescent fire-resistant paper sheets doped with different contents of Tb^{3+} ions; (d) photoluminescence emission spectra of the photoluminescent fire-resistant paper sheets doped with different contents of Eu^{3+} ions (the insets in (c) and (d) show the curves of the relationship between the luminescence intensity and the content of doped Ln^{3+} (Ln = Tb, Eu) ions, respectively). (Reprinted with permission from reference [63])

at an excitation wavelength of 378 nm (Figure 4.26c). In the emission spectrum of the photoluminescent fire-resistant paper doped with Eu^{3+} ions (Figure 4.26d), the characteristic peaks of Eu^{3+} ions were observed. The digital images (insets in Figure 4.26a and b) show the photoluminescence color of the photoluminescent fire-resistant paper sheets doped with Tb^{3+} and Eu^{3+} ions, respectively, under irradiation by a UV lamp (~365 nm), which showed a strong green and red color, respectively. The effect of the doped lanthanide ion concentration on the photoluminescence intensity of the paper was investigated. The experimental results indicated that the doping concentration of Ln^{3+} (Ln = Tb, Eu) ions had no obvious effect on the position of photoluminescence emission peak, but the emission peak intensity could be adjusted by regulating the doping concentration of rare earth ions. In Figure 4.26c, the photoluminescence intensity increased with increasing doping concentration of Tb^{3+} ions within a range below 5 mol% but dramatically decreased with further increasing doping concentration, which may due to concentration quenching effect. For the photoluminescence emission spectrum of the photoluminescent fire-resistant paper doped with Eu^{3+} ions, a similar phenomenon was observed, as shown in Figure 4.26d. The photoluminescence intensity increased with increasing concentrations of Eu^{3+} ions, reaching the maximum value at a doping concentration of 7.5 mol%, and then the intensity decreased at a doping concentration of 10 mol%.[63]

4.7. CATALYTIC FIRE-RESISTANT PAPER

Catalysts are widely used in chemical engineering and industrial production, and occupy an extremely important position in the modern chemical industry. In recent years, nanostructured catalysts have attracted widespread interest and attention. The nanostructured catalysts have small sizes, large specific surface areas, high surface activities, and excellent catalytic performance. However, a problem is that in a liquid-phase catalytic reaction system, when the catalytic reaction is completed, it is difficult to separate the nanostructured catalyst powder from the liquid-phase reaction system, which makes the recovery of the catalyst difficult and the recycling cost high. In addition, the catalyst nanoparticles are prone to severe agglomeration in the liquid-phase reaction system, resulting in a significant reduction in the catalytic performance of the catalyst.

The author's research group developed a new kind of layer-structured catalytic fire-resistant paper consisting of ultralong hydroxyapatite nanowires and gold nanoparticles, as shown in Figure 4.27.[64] A layered structure with interlayer spacing ranging from 0.5 μm to 3 μm could be observed on the cross-section of the as-prepared catalytic fire-resistant paper. Ultralong hydroxyapatite nanowires were synthesized by the calcium oleate precursor solvothermal method in a stainless steel autoclave with a volume of 10 L, which was a relatively large scale in the laboratory synthesis of nanostructured materials (Figure 4.27a). Gold nanoparticles were uniformly dispersed on ultralong hydroxyapatite nanowires in the catalytic fire-resistant paper, which could effectively prevent the agglomeration of gold nanoparticles and enhance the catalytic activity. In addition, ultralong hydroxyapatite nanowires interwove with each other to form a porous network and a layered structure, which could greatly increase the chance of contact with catalyst nanoparticles when the reactants flowed through the paper in the catalytic reaction. The layer-structured catalytic fire-resistant paper exhibited a purple color and high flexibility (Figure 4.27b and c). The catalytic fire-resistant paper could be easily separated and recycled from the liquid-phase reaction system. In order to investigate the catalytic performance, five pieces of catalytic fire-resistant paper sheets with a diameter of 2.5 cm were cut from a large-sized catalytic fire-resistant paper randomly, and the continuous flow catalytic activities were measured, and the experimental results are shown in Figure 4.27d. The experiments showed that the conversion efficiencies of all samples were higher than 99.0%, showing the uniform, stable, and high catalytic performance of

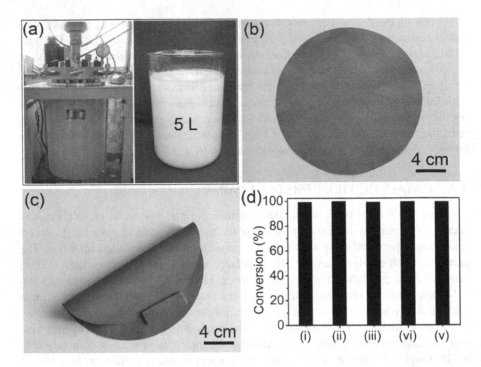

FIGURE 4.27 Digital images and continuous flow catalytic performance of the catalytic fire-resistant paper with a diameter of 20 cm. (a) The digital images of a stainless steel autoclave with a volume of 10 L for the scaled-up synthesis of ultralong hydroxyapatite nanowires by the calcium oleate precursor solvothermal method, and the obtained aqueous suspension containing ultralong hydroxyapatite nanowires after washing; (b and c) digital images of the highly flexible catalytic fire-resistant paper with a diameter of 20 cm; (d) conversion efficiencies of the reduction reaction of 4-nitrophenol to 4-aminophenol by sodium borohydride at room temperature using five pieces of small-sized catalytic fire-resistant paper sheets cut from a large-sized catalytic fire-resistant paper randomly. (Reprinted with permission from reference [64])

the large-sized catalytic fire-resistant paper. After the catalytic reaction was completed, as long as the catalytic fire-resistant paper was taken out of the solution, the nanostructured catalyst could be recovered and reused conveniently and quickly.

The fire-resistance tests of the catalytic fire-resistant paper in comparison with control samples were carried out, and the experimental results are shown in Figure 4.28. As expected, commercial filter paper consisting of plant fibers was completely burned to ashes in only 6 seconds (Figure 4.28a and d). In contrast, owing to the high thermal stability and excellent nonflammability of ultralong hydroxyapatite nanowires, the as-prepared catalytic fire-resistant paper exhibited excellent fire-resistant performance (Figure 4.28b and c). This unique feature of the catalytic fire-resistant paper allows it to be a promising nanostructured catalyst for high-temperature catalytic applications.[64]

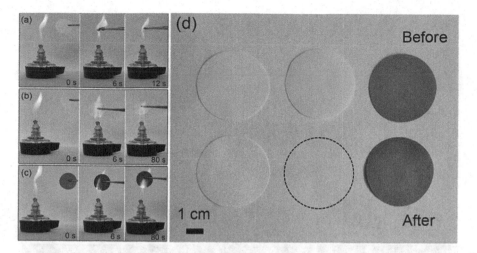

FIGURE 4.28 Fire-resistance tests of the catalytic fire-resistant paper. (a) The commercial filter paper was burned to ashes in several seconds; (b) the pure fire-resistant paper composed of only ultralong hydroxyapatite nanowires was nonflammable; (c) the layer-structured catalytic fire-resistant paper was also nonflammable; (d) digital images of the pure fire-resistant paper (left), commercial filter paper (middle), and catalytic fire-resistant paper (right) before and after the fire-resistance tests. (Reprinted with permission from reference [64])

4.8. PHOTOTHERMAL FIRE-RESISTANT PAPER

Photothermal materials have the ability to effectively absorb light and convert light into heat, and have important applications in various fields. The author's research group developed a highly flexible photothermal fire-resistant inorganic paper comprising a fire-resistant paper consisting of ultralong hydroxyapatite nanowires and a small amount of glass fibers (as the reinforcing material to enhance the mechanical strength of the paper) as the thermal insulation support and a layer of carbon nanotubes as the light absorber for solar energy–driven seawater desalination and water purification.[65] The as-prepared photothermal fire-resistant paper has merits such as high water evaporation efficiency, interconnected porous structure for continuous water transportation and evaporation, high thermal stability, good thermal insulation for heat localization, high stability during recycling, and long-time usage. The photothermal fire-resistant paper can be used to generate drinkable water from both real seawater and wastewater containing high concentrations of heavy metal ions, industrial dyes, and biological bacteria.

The solar energy–driven water evaporation using the as-prepared photothermal fire-resistant paper is schematically presented in Figure 4.29a. The photothermal fire-resistant paper with a porous structure could float spontaneously on the water surface, converting the absorbed solar energy into heat for water evaporation. The layer of carbon nanotubes of the photothermal fire-resistant paper was hydrophobic and could stay above the water surface, while the fire-resistant paper was hydrophilic and was immersed in water. The generated water vapor condensed into clean fresh

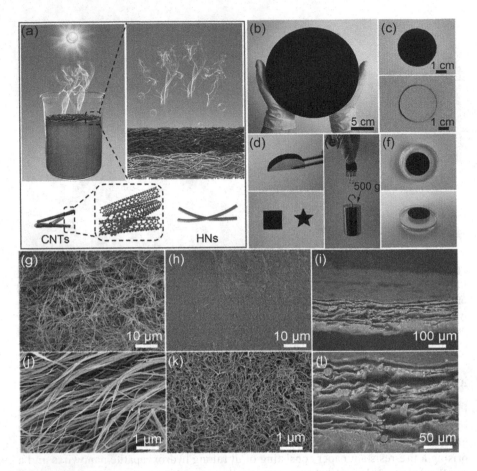

FIGURE 4.29 (a) Schematic illustration of solar energy–driven water evaporation using the as-prepared photothermal fire-resistant paper consisting of ultralong hydroxyapatite nanowires (HNs), glass fibers, and carbon nanotubes (CNTs); (b–f) digital images: (b) a photothermal fire-resistant paper sheet with a diameter of 20 cm, (c) the black top surface and the white bottom surface of a photothermal fire-resistant paper, (d) the photothermal fire-resistant paper exhibited a high flexibility and good processability, (e) four sheets of the photothermal fire-resistant paper could bear a weight of 500 g, (f) the photothermal fire-resistant paper could spontaneously float on the water surface; (g–l) SEM images: (g, j) the top surface morphology of the fire-resistant paper without a layer of carbon nanotubes, (h, k) the top surface morphology and (i, l) cross-section of the photothermal fire-resistant paper. (Reprinted with permission from reference [65])

water on a cold condenser. Figure 4.29b shows a photothermal fire-resistant paper with a diameter of 20 cm. In addition, the scaled-up production of the photothermal fire-resistant paper was realized employing a large stainless steel autoclave with a volume of 100 L for the large-scale synthesis of ultralong hydroxyapatite nanowires by the calcium oleate precursor solvothermal/hydrothermal method. Figure 4.29c shows digital images of the black upper surface and white bottom surface of the photothermal fire-resistant paper. The photothermal fire-resistant paper showed a high flexibility and good processability (Figure 4.29d). Owing to the nacre-like multilayered structure and "steel-reinforced concrete structure" constructed with ultralong hydroxyapatite nanowires and glass fibers, the photothermal fire-resistant paper exhibited good mechanical properties, and four sheets of the photothermal fire-resistant paper could bear a weight of 500 g, which was about 3,800 times its own weight (Figure 4.29e). In addition, the photothermal fire-resistant paper could float on the water surface (Figure 4.29f). The self-floating ability of the photothermal fire-resistant paper could enable the solar energy–converted heat to be localized at the water–air interface to inhibit heat loss. SEM images (Figure 4.29g, h, j, k) show that the top surface of both the fire-resistant paper without a layer of carbon nanotubes and the photothermal fire-resistant paper was highly porous. The photothermal fire-resistant paper had an average pore size of ~410.8 nm and a porosity of ~89.8%. The hydrophilic feature and interconnected porous structure could provide available pore channels for rapid water transportation. The upper carbon nanotube layer also exhibited a porous structure and could ensure the rapid water vapor escape. In addition, a layered structure was observed on the cross-section of the photothermal fire-resistant paper (Figure 4.29i, l).[65]

High thermal stability is important for solar energy–driven water evaporation, especially at high power densities and in high-temperature environments. The thermal stabilities of the as-prepared photothermal fire-resistant paper, and several commonly used thermal insulating materials such as the commercial filter paper, air-laid paper, polystyrene foam, and expandable polyethylene foam were evaluated, as shown in Figure 4.30. The thermogravimetric curves in Figure 4.30a and b show that the filter paper, air-laid paper, polystyrene foam, and expandable polyethylene foam decomposed at ~290°C, and almost completely decomposed at around 500°C. Owing to the high thermal stability of the fire-resistant inorganic paper consisting of ultralong hydroxyapatite nanowires and glass fibers without a layer of carbon nanotubes, only 0.5 wt.% loss was observed at the testing temperatures ranging from room temperature to 900°C in the thermogravimetric curve, which was attributed to the adsorbed water and oleic acid molecules. Carbon nanotubes also showed a high thermal stability below 400°C, and only a small amount of weight loss was observed at temperatures below 400°C. The weight loss of the photothermal fire-resistant paper was ~5 wt.% in the temperature range from room temperature to 900°C. Moreover, as shown in Figure 4.30b, the polystyrene foam and the expandable polyethylene foam obviously collapsed at temperatures above 120°C. The air-laid paper shrank at 120°C and severely curled at temperatures above 180°C. Furthermore, the filter paper showed crispation and an initial color change at 180°C and obvious color change at 240°C. By comparison, the fire-resistant inorganic paper without a layer of carbon nanotubes, and the photothermal fire-resistant paper exhibited no obvious change in shape, size, and color in the testing temperature range from 120°C to 400°C, indicating their high thermal stabilities. These merits of the photothermal fire-resistant paper enable its promising application in solar energy–driven photothermal water evaporation and water purification, especially in high-temperature environments.[65]

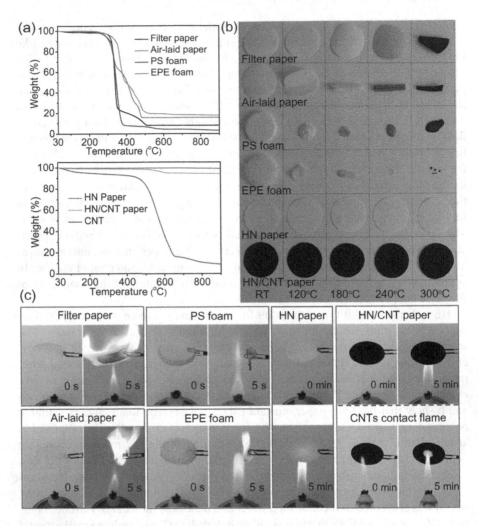

FIGURE 4.30 Thermal stability and fire-resistance tests. (a) Thermogravimetric curves of filter paper, air-laid paper, polystyrene (PS) foam, expandable polyethylene (EPE) foam, fire-resistant inorganic paper (HN) consisting of ultralong hydroxyapatite nanowires and glass fibers, photothermal fire-resistant paper (HN/CNT) consisting of ultralong hydroxyapatite nanowires, glass fibers, and carbon nanotubes; (b) digital images of the samples after thermal treatment at different temperatures ranging from 120 °C to 300°C for 30 minutes; (c) digital images of the samples before and after fire-resistance tests. The photothermal fire-resistant paper could be well preserved when the fire-resistant inorganic paper was in the alcohol flame, while the carbon nanotube component was removed when it was directly in the alcohol flame. (Reprinted with permission from reference [65])

Organic components are usually used as thermal insulating materials for solar energy–driven water evaporation, which are thermally unstable and highly flammable with potential risks. If the photothermal materials are nonflammable and have high thermal stability, advanced photothermal applications can be realized under extreme conditions. Figure 4.30c shows the fire-resistance tests in an alcohol flame of filter paper, air-laid paper, polystyrene foam, expandable polyethylene foam, fire-resistant inorganic paper without a layer of carbon nanotubes, and photothermal fire-resistant paper. The experiments showed that organic cellulose-based paper, such as the filter paper, and polymer-based materials, such as the air-laid paper, polystyrene foam, and expandable polyethylene foam, rapidly shrank and completely burned in the alcohol flame in seconds. In contrast, the fire-resistant inorganic paper and the photothermal fire-resistant paper could well preserve the shape, size, and initial color in the alcohol flame for more than 5 minutes, indicating their high thermal stability and excellent fire-resistant property. When the top carbon nanotube layer was exposed to the alcohol flame, the carbon nanotubes at the central area were removed in 5 minutes. The experimental results demonstrated that the photothermal fire-resistant paper possessed high thermal stability, and excellent fire resistance and thermal insulation.

The highly porous networked structure, excellent high-temperature stability, and fire resistance of the photothermal fire-resistant paper are favorable for the solar energy–driven water evaporation application. The thermal conductivities of the fire-resistant inorganic paper without a layer of carbon nanotubes and the photothermal fire-resistant paper were 0.085 $Wm^{-1}K^{-1}$ and 0.088 $Wm^{-1}K^{-1}$, respectively, which were significantly lower than that of water (0.6 $Wm^{-1}K^{-1}$) and those of many previous reported solar energy–driven water evaporation materials. The excellent thermal insulation of the photothermal fire-resistant paper could effectively minimize the heat loss from the paper surface to bulk water, which was desirable to realize high-performance solar energy–driven water evaporation. The light absorption and solar energy–driven water evaporation properties of the photothermal fire-resistant paper will be discussed in Section 5.5.3.

4.9. LIGHT-DRIVEN SELF-PROPELLED SUPERHYDROPHOBIC PHOTOTHERMAL FIRE-RESISTANT PAPER

In recent years, stimulus-response self-propelled materials have attracted widespread interest owing to their good application prospects in various fields, such as intelligent robots, microelectronics, environmental engineering, and biomedicine. Stimulus-responsive actuators that can respond to external stimuli and convert external energy into dynamic movement behaviors are highly desirable for many applications. However, many self-propelled materials reported in the literature have problems such as the slow moving velocity, long response time, complex preparation procedures, and poor environmental applicability. These shortcomings limit the practical applications of self-propelled materials. Among various driving schemes, the Marangoni propulsion produced by the photothermal effect is promising for enabling stimulus-responsive motion in a contactless and switchable manner. The Marangoni effect is considered as an attractive strategy for achieving self-propelled movement on the water surface because of its advantages. The localized change of temperature or chemical constituent can generate a surface tension gradient, and the liquid flows spontaneously from the area with a lower surface tension to the area with a higher surface tension, resulting in the macroscopic movement of the floating object.

Recently, the author's research group developed a highly flexible, superhydrophobic, and photothermal fire-resistant paper based on ultralong hydroxyapatite nanowires with the surface modification with both polydopamine (PDA) and oleylamine (OA) for controllable light-driven self-propelled motion on a liquid surface.[66] The highly stable superhydrophobicity could induce a thin air layer between the floating photothermal fire-resistant paper and the water surface to lower the fluid drag force. In addition, the photothermal effect could produce the Marangoni flow and controllable light-driven motion. The motional direction and velocity could be controlled by adjusting the position of the irradiated light and light power density. Furthermore, because of the high flexibility and excellent processability of the light-driven self-propelled superhydrophobic photothermal fire-resistant paper, various light-driven self-propelled actuators with designed shapes could be prepared through simple curling and cutting to achieve specific movement behaviors.[66]

As shown in Figure 4.31a, the as-prepared light-driven self-propelled superhydrophobic photothermal fire-resistant paper exhibited good mechanical properties. Five sheets of the light-driven self-propelled superhydrophobic photothermal fire-resistant paper could bear a weight of 500 g, which was ~4200 times greater than the

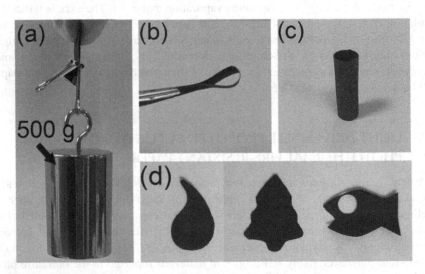

FIGURE 4.31 Characterization of the as-prepared light-driven self-propelled superhydrophobic photothermal fire-resistant paper. (a) Five sheets of the paper could bear a weight of 500 g; (b) the paper exhibited a high flexibility; (c, d) the paper showed excellent processability. (Reprinted with permission from reference [66])

weight of the paper. Furthermore, the light-driven self-propelled superhydrophobic photothermal fire-resistant paper had a high flexibility and excellent processability, and it could be bent at arbitrary angles (Figure 4.31b), and could be processed into desired two- and three-dimensional shapes by simple curling and cutting processes (Figure 4.31c and d). The light-driven self-propelled superhydrophobic photothermal fire-resistant paper also showed excellent superhydrophobicity, and a silvery sheen was observed when it was immersed in water, indicating that a thin layer of air was formed between the paper and water.

The light-driven self-propelled superhydrophobic photothermal fire-resistant paper possessed excellent photothermal conversion performance and highly stable superhydrophobic property, which could enable on-demand motion under the manipulation of the irradiated light, as shown in Figure 4.32a. The photothermal effect of the light-driven self-propelled superhydrophobic photothermal fire-resistant paper could absorb light and convert it into heat, and then the heat could be transmitted to the surrounding water surface, generating the temperature gradient, Marangoni flow, and motion of the paper. The superhydrophobic property could enable the paper to float on the water surface and form a thin layer of air between the paper and the water surface that could effectively reduce the fluidic drag force during the movement process on the water surface. One of the merits is that the moving direction and velocity of the paper could be manipulated by the irradiated light to achieve on-demand motions. As shown in Figure 4.32b, the moving direction of a small piece of the light-driven self-propelled superhydrophobic photothermal fire-resistant paper could be controlled by changing the light irradiation position, leading to various movement behaviors of the paper on the water surface, including going forward, going backward, turning left, and turning right.[66]

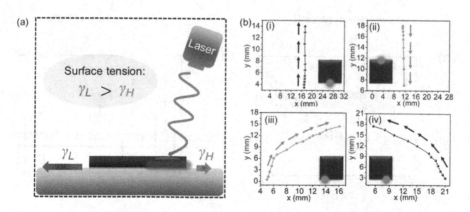

FIGURE 4.32 The movement performance of the as-prepared light-driven self-propelled superhydrophobic photothermal fire-resistant paper. (a) Schematic illustration of the controllable motion triggered by the Marangoni effect; (b) moving tracks of the paper: (i) going forward, (ii) going backward, (iii) turning right, and (iv) turning left (the insets show the corresponding light-irradiated positions on the paper). (Reprinted with permission from reference [66])

The moving velocity and response time of the light-driven self-propelled super-hydrophobic photothermal fire-resistant paper could be controlled by the irradiated light power density, as shown in Figure 4.33a and b. The average velocity of the paper increased from 2.62 ± 1.09 mm s^{-1} to 22.14 ± 4.89 mm s^{-1}, and the response time decreased from 2.16 ± 0.56 seconds to 0.10 ± 0.03 seconds by increasing the power density of 808 nm near infrared light from 1 W cm^{-2} to 6 W cm^{-2}. In addition, the light-driven self-propelled superhydrophobic photothermal fire-resistant paper could move under various conditions (for example, at a relative low temperature (4°C), 0.1 M sodium chloride solution, and acidic and alkaline conditions) (Figure 4.33c), and the average velocity had only a small change under these conditions (Figure 4.33d). Furthermore, the light-driven self-propelled superhydrophobic photothermal fire-resistant paper exhibited a good load-bearing capacity, and the paper could float on the water surface even with a load of ~80 times its own weight. Under the irradiation of near infrared light (2 W cm^{-2}), the average velocity of the light-driven self-propelled superhydrophobic photothermal fire-resistant paper with a load of ~10 times its own weight reached 5.05 mm $s^{-1} \pm 1.62$ mm s^{-1}, which was slightly lower than that of the paper without a load.

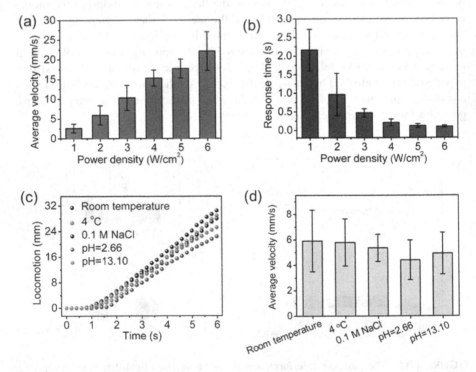

FIGURE 4.33 The movement properties of the light-driven self-propelled superhydrophobic photothermal fire-resistant paper. (a) Average velocity and (b) response time of the paper irradiated by near infrared light with different power densities; (c) locomotion plots and (d) average velocity of the paper under different conditions. (Reprinted with permission from reference [66])

The light-driven self-propelled superhydrophobic photothermal fire-resistant paper possessed excellent processability, and various actuators with complex shapes could be prepared by simple curling and cutting procedures. As shown in Figure 4.34a and b, two types of light-driven self-propelled actuators with designed shapes were prepared using the light-driven self-propelled superhydrophobic photothermal fire-resistant paper by a simple cutting process. The well-designed shapes of the actuators could achieve specific movement behaviors with good controllability. For instance, an arrow-shaped actuator could be accurately manipulated by the irradiated near infrared light and move along a specific and preplanned path (Figure 4.34c). Moreover, the rotatory motion was realized by a cross-shaped actuator. Driven by the irradiated near infrared light with a power density of 2 W cm^{-2}, the cross-shaped actuator could rotate clockwise with an angular velocity of 0.52 rad s^{-1} (Figure 4.34d). Compared with other techniques used for constructing actuators with specific shapes, the light-driven self-propelled superhydrophobic photothermal fire-resistant paper was able to construct light-driven functional actuators in a low-cost, quick, and efficient manner.[66]

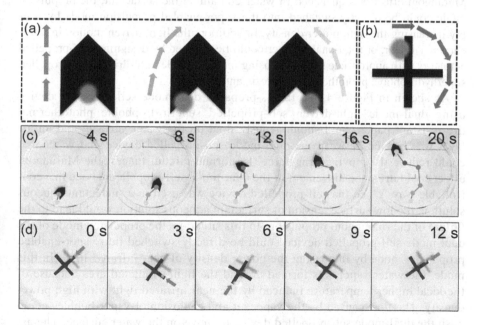

FIGURE 4.34 The moving behaviors of actuators with different shapes made from the light-driven self-propelled superhydrophobic photothermal fire-resistant paper. (a) An arrow-shaped actuator and its various moving directions controlled by irradiating near infrared light at specific positions; (b) a cross-shaped actuator and its light-driven rotatory movement; (c) an arrow-shaped actuator moved along a specific and preplanned path under the control of the irradiating near infrared light; (d) a cross-shaped actuator rotated clockwise driven by the irradiating near infrared light. (Reprinted with permission from reference [66])

The direct conversion of external energy into mechanical work by the self-propelled motors inspires the development of miniaturized machines. Although many previous studies demonstrated that the Marangoni propulsion induced by the photothermal effect is a feasible strategy to achieve direct light-to-work transformation, a weakness of this strategy is that the Marangoni effect is quenched in the presence of surfactants, which restrict the applications of the Marangoni effect-based self-propelled devices in the case of surfactant solutions.[67]

Several strategies were reported to realize the self-driven motion, but in some cases multiple power sources were necessary, leading to complicated operation in diverse environments. Therefore, the dual-mode self-propelled system based on a single power source is highly desirable. The author's research group developed the single-light-actuated dual-mode propulsion at the liquid–air interface using flexible, superhydrophobic, and thermostable photothermal fire-resistant paper made from ultralong hydroxyapatite nanowires, titanium sesquioxide (Ti_2O_3) particles, and poly(dimethylsiloxane) coating.[67] The superhydrophobic surface could enable the thermostable photothermal paper to float on the water surface spontaneously and significantly reduce the water drag force. In the usual situation, the heat power produced by the photothermal effect could trigger the Marangoni propulsion. While the Marangoni effect was quenched in water containing the surfactant, the propulsion mode could be directly switched into the vapor-enabled propulsion mode by simply increasing the light power density. In addition, the light-driven motion in a linear, curvilinear, or rotational manner could be realized by designing self-propelled actuators with appropriate shapes by using the dual-mode light-driven self-propelled superhydrophobic photothermal fire-resistant paper.

As shown in Figure 4.35a, the as-prepared dual-mode self-propelled device using dual-mode light-driven self-propelled superhydrophobic photothermal fire-resistant paper could spontaneously float on the water surface, and an air layer was formed between the dual-mode self-propelled device and water that could reduce the moving drag force. In normal circumstances, the Marangoni effect could drive the movement of the dual-mode self-propelled device in a controllable way. When the self-propelled device was used in a surfactant aqueous solution, the low surface tension of surfactant aqueous solution would lead to the failure of the Marangoni propulsion. In this situation, the propulsion mode of the dual-mode self-propelled device could be directly switched into vapor-enabled propulsion mode by increasing the power density of near infrared light. In this mode, the water vapor was formed around the light-irradiated area because of the local higher temperature induced by the near infrared light with high power density. The flow caused by the vapor jet and explosion of vapor bubbles could push the dual-mode self-propelled device to move on the water surface. The as-prepared dual-mode light-driven self-propelled superhydrophobic photothermal fire-resistant paper exhibited high flexibility and excellent processability (Figure 4.35b), and different devices with complex shapes could be prepared via simple cutting, folding, and curling processes (Figure 4.35c and d). When the dual-mode light-driven self-propelled superhydrophobic photothermal fire-resistant paper was soaked in water, a silvery sheen was observed on the surface of the

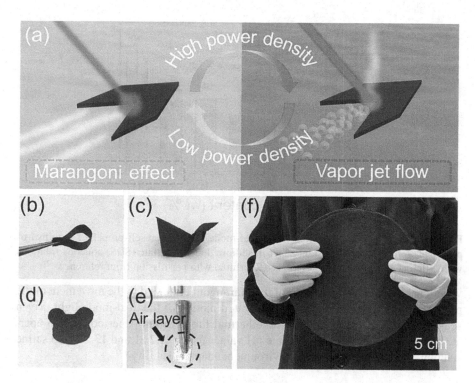

FIGURE 4.35 Single-light-operated dual-mode self-propelled propulsion at the liquid/air interface and the dual-mode light-driven self-propelled superhydrophobic photothermal fire-resistant paper. (a) Schematic illustration of the single-light-operated device for dual-mode self-propelled movement; (b–d) the dual-mode light-driven self-propelled superhydrophobic photothermal fire-resistant paper showed a high flexibility and good processability; (e) the dual-mode light-driven self-propelled superhydrophobic photothermal fire-resistant paper was submerged in water and an air layer was formed on the surface; (f) a dual-mode light-driven self-propelled superhydrophobic photothermal fire-resistant paper with a diameter of 20 cm. (Reprinted with permission from reference [67])

paper (Figure 4.35e). Because of the superhydrophobic property of the dual-mode light-driven self-propelled superhydrophobic photothermal fire-resistant paper, air bubbles were trapped on its superhydrophobic surface to form a thin air layer between the paper and water, and the air layer could reduce the moving drag force. Figure 4.35f shows a piece of dual-mode light-driven self-propelled superhydrophobic photothermal fire-resistant paper with a diameter of 20 cm.

The propulsion performance of the dual-mode light-driven self-propelled superhydrophobic photothermal fire-resistant paper in the surfactant aqueous solution was investigated. As shown in Figure 4.36, under the irradiation of near infrared light with a power density of 12 W cm^{-2}, the average moving velocity of the paper increased from 1.65 ± 1.14 mm s^{-1} to 11.89 ± 2.39 mm s^{-1} with increasing surfactant content from 0.5 wt.% to 10 wt.% because of the decreased surface tension of water. Because the increased surfactant concentration weakened the Marangoni effect, the

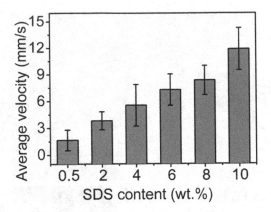

FIGURE 4.36 Average velocities of the dual-mode light-driven self-propelled superhydrophobic photothermal fire-resistant paper in sodium dodecyl sulfate (SDS) aqueous solutions with different surfactant concentrations. (Reprinted with permission from reference [67])

vapor-enabled propulsion instead of the Marangoni effect led to the movement of the paper. Furthermore, the single-light-operated motion of the dual-mode light-driven self-propelled superhydrophobic photothermal fire-resistant paper could also operate under conditions such as extreme pH values (e.g., pH 2.31 and 13.22) and saline solution, indicating its excellent environment adaptable ability.[67]

5 Applications of the Fire-Resistant Paper

The new kind of the fire-resistant paper based on ultralong hydroxyapatite nanowires has many advantages, such as high flexibility, high biocompatibility, high-quality natural white color without bleaching, high surface smoothness without surface sizing, environment friendliness, excellent resistance to both fire and high temperatures, etc. The fire-resistant paper can be used as the cultural paper for writing and printing, and is expected to be used for long-term safe preservation of important documents, archives, and books. In addition, the fire-resistant paper also has promising applications in many fields such as biomedical, environmental, energy, electronics, and information technology, in which traditional plant fiber paper cannot be used. The ultralong hydroxyapatite nanowires, which are the raw material of the new type of fire-resistant paper, can be synthesized artificially using common chemicals without consuming valuable natural resources such as trees. The whole manufacturing process of the fire-resistant paper is environment friendly and will not pollute the environment. It is expected that the new fire-resistant paper has a promising prospect for commercialization and large-scale applications.

5.1. FIRE-RESISTANT PAPER AS A SPECIALTY PAPER

5.1.1. PROTECTION AND PERMANENT SAFE PRESERVATION OF IMPORTANT DOCUMENTS, ARCHIVES, AND BOOKS

In contrast to flammable traditional paper based on plant fibers, the fire-resistant paper based on ultralong hydroxyapatite nanowires is nonflammable and high-temperature resistant, and can be used for writing and printing. Therefore, the fire-resistant paper can be used for making fireproof paper–based books, documents, archives, etc. In this way, the protection and permanent safe preservation of important documents, archives, and books can be realized using the fire-resistant paper.

In addition to excellent resistance to both fire and high temperature, the fire-resistant paper based on ultralong hydroxyapatite nanowires has excellent heat insulation performance, as shown in Figure 3.5. Therefore, the fire-resistant paper can be used as the nonflammable cover for traditional flammable paper–based books, documents, archives, etc., to protect these flammable paper products from being ruined by fire and high temperatures.

5.1.2. FIRE-RESISTANT "XUAN PAPER" FOR CALLIGRAPHY AND PAINTING

Among various kinds of paper, Xuan paper is an excellent representative of traditional handmade papers. Xuan paper originated from Jing County, Anhui Province, in China. Xuan paper was first mentioned in an ancient Chinese book entitled "Notes of Past Famous Paintings" by Yanyuan Zhang, a famous theorist on calligraphy and painting in the Tang Dynasty, China in the 9th century.[68] In 2009, the traditional handicraft of making Xuan paper was inscribed on the Representative List of the Intangible Cultural Heritage of Humanity by the Educational, Scientific and Cultural Organization of the United Nations.[69]

Xuan paper has advantages such as excellent durability, ink wetting, wet deformation, and resistance to insects and mildew, and it is the most durable paper in the world and enjoys a great reputation as "the king of paper that lasts for a 1,000 years." Xuan paper is the best material carrier for calligraphy and paintings. Many famous ancient calligraphy and painting works and books using Xuan paper have been well preserved and survive even today. According to the statistics of the National Library of China, there are about 30 million ancient books in China; and the majority were printed using Xuan paper and have been well preserved.[70] The excellent durability of Xuan paper is attributed to its unique raw materials and the handmade manufacturing process, which involves more than 100 steps. The bark of *Pteroceltis tatarinowii*, a common species of elm in the area, is used as the main raw material of Xuan paper. The bark fibers of *Pteroceltis tatarinowii* have good flexibility and a high degree of evenness. In addition, limestone particles are deposited on the surface of *Pteroceltis* bark fibers, which can neutralize the acidic substances produced by the hydrolysis of cellulose fibers and those present in the environment, and this is one of the important reasons for the durability of Xuan paper.[71] Another reason for the durability of Xuan paper is its complex handmade production process under mild treatment conditions including more than 100 steps, which minimize chemical damages to the plant fibers.[68, 71]

Xuan paper is mainly classified into unprocessed Xuan paper and processed Xuan paper. The unprocessed Xuan paper is prepared without special post-processing and has high water absorption, which causes the ink to blur on it, making it suitable for freehand brushwork paintings. Processed Xuan paper is prepared by treating the unprocessed Xuan paper with bone glue and potassium aluminum sulfate solution, resulting in a hydrophobic surface with a reduced capacity to absorb ink, and it is suitable for the meticulous paintings which require artists to render delicate details to drawings.[72]

Although traditional Xuan paper has many advantages, it has problems which need to be solved. For example: (1) the raw materials of Xuan paper are produced from Jing County, Anhui Province, in China which are severely scarce; (2) the traditional handmade production process of Xuan paper needs an ultralong production cycle of about two years, leading to low output and high cost; (3) the organic origin of Xuan paper usually results in the paper degradation, yellowing, and deteriorating during the long-term natural aging process; (4) the main weakness of the traditional Xuan paper is its high flammability—numerous precious calligraphy and painting works and books were burned to ashes in fire in the past centuries.[72]

The author's research group developed a new kind of highly flexible inorganic fire-resistant analogous Xuan paper based on ultralong hydroxyapatite nanowires with high whiteness, excellent resistance to fire and high temperatures, superdurability of thousands of years, unique ink wetting, and excellent anti-mildew benefiting from its 100% inorganic origin, which are far superior over traditional Xuan paper.[72] The inorganic fire-resistant analogous Xuan paper can be well preserved with no obvious deterioration in properties even after simulated aging for 3,000 years. The most attractive merits of the inorganic fire-resistant analogous Xuan paper are its excellent nonflammability and high thermal stability, which can well safeguard precious calligraphy and painting works, documents, and books for a very long period of time without the fear of being destroyed by fire or yellowing. The production process of the inorganic fire-resistant analogous Xuan paper is environment friendly and simple with only a few steps in contrast to more than 100 steps for traditional Xuan paper, and it is highly efficient and takes only 2–3 days instead of about two years for traditional Xuan paper. The future commercialization of the inorganic fire-resistant analogous Xuan paper will solve the problems of traditional Xuan paper, such as a severe shortage of geography-specific raw materials, ultralong production cycle, yellowing, and high flammability, and extend its applications to fire-resistant books, documents, and archives in addition to calligraphy works and paintings.

The unique structure of the inorganic fire-resistant analogous Xuan paper with high flexibility was innovatively designed based on ultralong hydroxyapatite nanowires as the main building material, glass fibers as the reinforcing material, and inorganic adhesive as the glue, which is similar to the "steel-reinforced concrete structure" in tall buildings. The design idea was based on the following considerations: (1) glass fibers with micrometer-sized diameters were adopted as the reinforcing framework material, similar to supporting steel rebars in tall buildings; (2) ultralong hydroxyapatite nanowires were used as the main building material, similar to the concrete in tall buildings; (3) the aspect ratios of ultralong hydroxyapatite nanowires (up to >10,000) were much higher than those of the plant fibers and flexible enough to interweave with each other to form a porous network structure; (4) ultralong hydroxyapatite nanowires had a relatively high specific surface area ($\sim$67.3 m^2 g^{-1}), which was higher than those of micrometer-sized cellulose fibers; (5) high specific surface area and unique structure were beneficial to increasing the contact points and binding force between the ultralong hydroxyapatite nanowires and also the mechanical strength of the inorganic fire-resistant analogous Xuan paper; (6) more importantly, ultralong hydroxyapatite nanowires were environment friendly, with high biocompatibility and high-quality whiteness (bleaching was not needed) and high thermal stability; (7) amorphous nanophase inorganic adhesive was designed, prepared, and used as the binder in the inorganic fire-resistant analogous Xuan paper. SEM and energy-dispersive X-ray spectroscopy analyses showed that the as-prepared inorganic adhesive consisted of amorphous nanoparticles, and included the elements of Si, K, Al, B, O, and P with a high specific surface area ($\sim$142.7 m^2 g^{-1}), which could further increase the mechanical strength of the inorganic fire-resistant analogous Xuan paper. The thermogravimetric and differential scanning calorimetry curves showed that the inorganic adhesive possessed a high thermal stability at temperatures up to 1,200°C. As an example, the weight ratio of three components in a typical sample of the inorganic fire-resistant analogous Xuan paper was as follows: ultralong hydroxyapatite nanowires/glass fibers/inorganic adhesive $\approx$ 68:17:15.

In the preparation process of the inorganic fire-resistant analogous Xuan paper, ultrasound was used to disperse glass fibers and wrap them with ultralong hydroxyapatite nanowires. SEM images showed that ultralong hydroxyapatite nanowires were adsorbed on the surface of glass fibers via the electrostatic interaction and van der Waals force to form a stable nanocomposite aqueous suspension, and then the inorganic adhesive was added under ultrasonic irradiation. The resulting aqueous suspension was filtered under vacuum to form a wet paper, which was pressed at a pressure of 4 MPa and dried at 105°C for 3 minutes. The production process of the inorganic fire-resistant analogous Xuan paper was environment friendly and simple, with only a few steps, including one-step synthesis of ultralong hydroxyapatite nanowires, washing, slurrying, paper forming, pressing, and drying, in contrast to more than 100 steps for the traditional Xuan paper. More importantly, the preparation of the inorganic fire-resistant analogous Xuan paper was highly efficient, and it took only 2–3 days, in contrast to about two years for traditional Xuan paper.[72] The inorganic fire-resistant analogous Xuan paper sheets with a diameter of 20 cm and in A3 size (297 mm × 420 mm) were prepared in the author's laboratory, as shown in Figure 5.1a and b. The as-prepared inorganic fire-resistant analogous Xuan paper was highly flexible with a high whiteness (Figure 5.1b and c).

In calligraphy and painting, the artists use flexible writing brushes to create expressive strokes and drawings, and the ink wetting performance on Xuan paper plays an important role in this process. The ink wetting behavior of the inorganic fire-resistant analogous Xuan paper was investigated, and the commercial unprocessed and processed Xuan paper sheets were used for comparison. The experiments indicated that the ink droplet on the commercial unprocessed Xuan paper spread rapidly and permeated easily into the paper, and the ink even went through the paper and was obviously visible on the back side of the paper, and the edges of ink and strokes were rough with burrs, as shown in Figure 5.2a. The ink droplet on the commercial processed Xuan paper exhibited a spherical shape with a little diffusion, and the edges of the ink and strokes were relatively smooth (Figure 5.2b). In comparison, the ink wetting behavior on the inorganic fire-resistant analogous Xuan paper was unique and different from those on the two kinds of the traditional Xuan paper. The ink droplet spread slightly on the inorganic fire-resistant analogous Xuan paper but did not penetrate through the paper, and no ink was observed on the back side of the paper, and the edges of ink and strokes were smooth (Figure 5.2c). In order to further study the effect of ink wetting, the ink was diluted to 2% and was dropped on the edge of a thick ink droplet. It was observed (Figure 5.2a) that a drop of the thick ink and a drop of the diluted ink on commercial unprocessed Xuan paper were well separated from each other with a slight diffusion on the interfacial area. When a drop of the diluted ink was dropped on the edge of a thick ink droplet on commercial processed Xuan paper, a rapid diffusion from the thick ink to diluted ink was observed, and then a uniform ink was formed. The inorganic fire-resistant analogous Xuan paper exhibited a similar ink wetting performance to that of commercial processed Xuan paper except that the ink droplet slightly diffused on the paper (Figure 5.2c). The surface hydrophilic/hydrophobic property obviously affected the ink wetting performance of the paper. Commercial unprocessed Xuan paper was

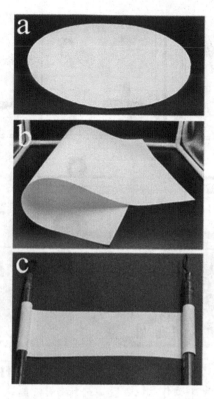

FIGURE 5.1 Digital images of the as-prepared inorganic fire-resistant analogous Xuan paper. (Reprinted with permission from reference [72])

superhydrophilic with a water contact angle of 0° and also an ink contact angle of 0°; the water droplet or ink droplet could diffuse rapidly on the commercial unprocessed Xuan paper, and the diffusion of both water and ink constituents was simultaneous on commercial unprocessed Xuan paper (Figure 5.2d). The commercial processed Xuan paper was hydrophobic with a water contact angle of 130.5° and an ink contact angle of 88.3°, and a water droplet or ink droplet exhibited a nearly spherical shape with little diffusion on commercial processed Xuan paper (Figure 5.2e). However, the inorganic fire-resistant analogous Xuan paper was superhydrophilic with a water contact angle of 0° and an ink contact angle of 36.9° (Figure 5.2f), and the diffusion of water and ink constituents was not simultaneous, and the water in the ink diffused more rapidly than the ink constituents. This unique phenomenon could lead to faster drying of the ink on the inorganic fire-resistant analogous Xuan paper than on traditional processed Xuan paper.[72]

SEM images (Figure 5.2h and i) indicate the porous structure of the traditional Xuan paper formed by the plant fibers. In comparison, an ordered texture of the inorganic fire-resistant analogous Xuan paper was observed at a low magnification because of the textured pattern of the cloth used in the preparation process (Figure 5.2j), and it showed a porous networked structure formed by ultralong hydroxyapatite

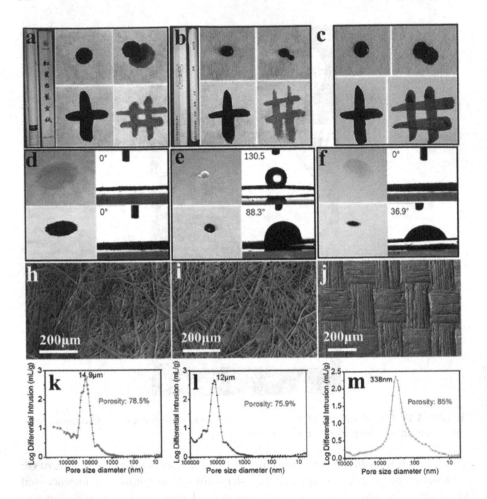

FIGURE 5.2 The ink wetting properties of the inorganic fire-resistant analogous Xuan paper in comparison with the commercial Xuan paper (a, d, h, k); commercial unprocessed Xuan paper (b, e, i, l); and commercial processed Xuan paper (c, f, j, m); inorganic fire-resistant analogous Xuan paper (c, f, j, m). (a–c) Digital images; (d–f) water contact angles; (h–j) SEM images; and (k–m) pore size distributions. (Reprinted with permission from reference [72])

nanowires. The most probable pore size (14.9 μm) and porosity (78.5%) of commercial unprocessed Xuan paper were similar to those (12 μm, 75.9%) of commercial processed Xuan paper (Figure 5.2k and l). In contrast, the inorganic fire-resistant analogous Xuan paper possessed a much smaller pore size (338 nm) and higher porosity (85%) (Figure 5.2m). Furthermore, the surface roughness of traditional Xuan paper was higher than that of the inorganic fire-resistant analogous Xuan paper. Higher roughness, larger pores, and hydrophilicity of commercial unprocessed Xuan paper caused the disordering of ink diffusion pathways and rough edges of the ink and strokes with obvious burrs, and the ink could penetrate through the paper. In contrast, the pores of the inorganic fire-resistant analogous Xuan paper were more uniform and smaller, and the paper surface was smoother, leading to the relatively uniform ink diffusion in each direction and smooth edges of the ink and strokes without obvious burrs (Figure 5.2c).

Mold is an important factor for the deterioration of traditional Xuan paper. The prevention of mold growth on the paper is vital for its long-term safe preservation. In order to study the anti-mildew properties of the inorganic fire-resistant analogous Xuan paper, two groups of experiments were carried out. In the first group of experiments, mold spores were sprayed on three kinds of paper sheets, and then the paper sheets were placed into an incubator to culture for 28 days at a constant temperature of $30 \pm 1°C$ and a humidity of 90%. The experimental results showed that no mold propagation was observed on the three kinds of paper sheets, indicating their good anti-mildew performance in the absence of external nutrients. However, in a control experiment without any paper sample, mold spores sprayed on the Bengal culture medium could grow and spread quickly, as shown in Figure 5.3a.

During the normal storage of calligraphy works and paintings, the surface of Xuan paper is easily polluted by the organic pollutants in air. Therefore, a second group of experiments were performed, in which the paper was cut into a square shape with a size of 3 cm × 3 cm and placed in Bengal culture medium containing mold spores and cultured for four days at a constant temperature of $30 \pm 1°C$ and a humidity of 90% (Figure 5.3b). The experiments indicated that three kinds of mold spores could not breed and spread on the inorganic fire-resistant analogous Xuan paper, and it could preserve a clean surface without the growth of mold, indicating the excellent anti-mildew performance of the inorganic fire-resistant analogous Xuan paper even in the presence of external nutrients. On the contrary, mold could grow and spread on both commercial unprocessed and processed Xuan paper sheets in Bengal culture medium, indicating that the anti-mildew performance of the traditional Xuan paper was not satisfactory in the presence of external nutrients. The composition of the traditional Xuan paper is mainly plant cellulose fibers, which can feed mold reproduction. In addition, an enzyme could be produced during the process of mold reproduction which could chemically attack and cleave cellulose macromolecules. Furthermore, the metabolism of mold could produce the acidic waste and further degrade the traditional Xuan paper. This is also one of the main reasons for the significantly degraded properties of the traditional Xuan paper during the long-term preservation process.[72]

Traditional Xuan paper is highly flammable due to its organic composition of plant fibers. In history, numerous precious calligraphy and painting works, documents, and books were burned to ashes in fire. The inorganic fire-resistant analogous Xuan paper is nonflammable and can well safeguard precious paper-based books, documents, archives, paintings, and calligraphy works for a very long period of time without the fear of being destroyed by fire and high temperatures. Figure 5.4 shows a comparison of the fire-resistant performance and high-temperature stability of three kinds of paper sheets. Traditional unprocessed Xuan paper and processed Xuan paper were highly flammable and burned to ashes in seconds in fire (Figure 5.4a). In contrast, the inorganic fire-resistant analogous Xuan paper

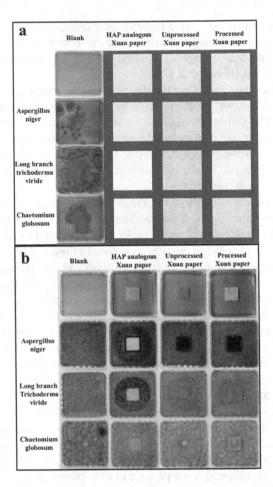

FIGURE 5.3 Anti-mildew performance of the inorganic fire-resistant analogous Xuan paper compared with the commercial Xuan paper. (a) Digital images of mold spores sprayed on three kinds of paper sheets and cultured for 28 days at 30 ± 1°C and a humidity of 90%; (b) digital images of three kinds of paper sheets immersed in the Bengal culture medium containing mold spores and cultured for four days at 30 ± 1°C and a humidity of 90%. (Reprinted with permission from reference [72])

exhibited excellent fire-resistant performance; it was nonflammable even if it was heated in fire for a long period of time (Figure 5.4a). The thermal stability tests indicated that both commercial unprocessed and processed Xuan paper turned yellow at 200°C and heavily carbonized at 400°C for 30 minutes. However, the inorganic fire-resistant analogous Xuan paper had no obvious change in both color and dimension at temperatures ranging from 200°C to 1,000°C (Figure 5.4b). Although the tensile strength of the inorganic fire-resistant analogous Xuan paper decreased after heat treatment at high temperatures, a tensile strength retention rate of 98.4% was achieved after heat treatment at 200°C for 30 minutes, and the tensile strength

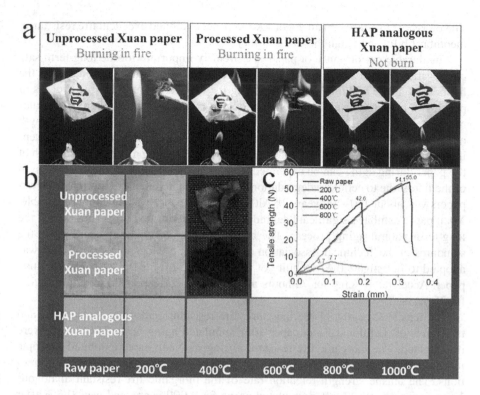

FIGURE 5.4 Resistance tests to fire and high temperatures of the inorganic fire-resistant analogous Xuan paper and commercial Xuan paper. (a) Fire-resistance tests of three kinds of paper sheets; (b) thermal stability tests of three kinds of paper sheets at different temperatures (200°C–1,000°C); (c) tensile strength versus strain of the inorganic fire-resistant analogous Xuan paper after heat treatment for 30 minutes at different temperatures (200°C–800°C). (Reprinted with permission from reference [72])

retention rate was 77.5% after heat treatment at 400°C for 30 minutes (Figure 5.4c). In comparison, traditional Xuan paper was heavily carbonized to ashes at 400°C for 30 minutes (Figure 5.4b).[72]

Cone calorimetry test has been recognized as one of the most acceptable fire testing methods that can provide information on materials' reaction to fire, and can provide the thermal parameters of materials, including the heat release rate (HRR), total heat release (THR), smoke produce rate (SPR), total smoke release (TSR), fire growth index (FGI), etc. Cone calorimetry tests were carried out to study the fire-resistance properties of the inorganic fire-resistant analogous Xuan paper. The experiments indicated that the inorganic fire-resistant analogous Xuan paper was well preserved after the testing, and no obvious damage was observed. The experimental results showed that the effective heat of combustion, mass loss rate, and heat release rate were essentially zero throughout the testing process, which can be explained by the fact that the inorganic fire-resistant analogous Xuan paper had the entire inorganic origin, and the paper could not be ignited throughout the whole heating process.

The inorganic fire-resistant analogous Xuan paper exhibited excellent fire resistance, incombustibility, and high thermal stability.

The durable performance of paper is extremely important for the long-term safe preservation of books, documents, archives, etc., which is mainly dependent on the fibers used as the building material. During the long-term natural aging process of paper, two most intuitive phenomena are decreased mechanical strength and whiteness. The mechanical strength of Xuan paper is determined by the intrinsic strength of the fibers and the bonding strength between the fibers. In addition to the weakening of the bonding strength between the fibers, the decreasing mechanical strength of the paper during aging can be explained mainly by the decreasing intrinsic strength of the fibers due to cellulose degradation. It was estimated that the lifetime of newspapers was about 150–300 years.[73] Although Xuan paper was much more durable, Xuan paper exhibited significant degradation and deteriorated properties during the long-term natural aging process. The accelerated aging method according to the standards of the Technical Association of the Pulp and Paper Industry of USA was adopted to investigate the durability of the inorganic fire-resistant analogous Xuan paper. Accelerated aging for 72 hours at $105 \pm 2°C$ corresponded to natural aging for 25 years.[74]

The tensile strengths of the inorganic fire-resistant analogous Xuan paper and two kinds of traditional Xuan paper after simulated aging up to 3,000 years were investigated. The inorganic fire-resistant analogous Xuan paper exhibited isotropic mechanical properties, and the tensile strength was the same along different directions. The tensile strength retention rate of the inorganic fire-resistant analogous Xuan paper was 95.2% after simulated aging for 2,000 years, and was 81.3% after simulated aging for 3,000 years. The tensile strength of the traditional Xuan paper usually depends on the orientation along the longitudinal direction (LD) or transverse direction (TD). The tensile strength retention rate of commercial unprocessed Xuan paper was only 61.2% (TD) and 48.6% (LD) after simulated aging for 2,000 years, and was only 38% (TD) and 42.8% (LD) after simulated aging for 3,000 years. The tensile strengths of commercial processed Xuan paper were lower than those of commercial unprocessed Xuan paper before simulated aging. For commercial processed Xuan paper, the tensile strength retention rate was 56.7% (TD) and 81.5% (LD) after simulated aging for 2,000 years, and 52.7% (TD) and 58.9% (LD) after simulated aging for 3,000 years. Comparing the three kinds of paper sheets after simulated aging for 2,000 years, the inorganic fire-resistant analogous Xuan paper exhibited a much higher tensile strength retention rate (95.2%) than commercial unprocessed Xuan paper (average 54.9%) and processed Xuan paper (average 69.1%). The tensile strength retention rate for the inorganic fire-resistant analogous Xuan paper was 81.3% after simulated aging for 3,000 years, while, on average, that of commercial unprocessed Xuan paper was 40.4% and that of processed Xuan paper was 55.8%.

The above experimental results can be explained by different building constituents of the three different kinds of paper sheets. The inorganic fire-resistant analogous Xuan paper consists of 100% inorganic constituents, which are highly stable during a long period of aging time. In contrast, traditional Xuan paper is made from plant cellulose fibers. Cellulose is an organic compound with the formula $(C_6H_{10}O_5)_n$, a polysaccharide consisting of a linear chain of several hundred to thousands of $\beta(1-4)$ linked D-glucose units which are not stable during a long period of aging.

SEM images showed that degradation was observed on the surface of commercial unprocessed Xuan paper after simulated aging for 2,000 years, and obvious holes formed on the paper surface, cellulose hydrolyzed to form granules in the holes, and cracks were observed on the cellulose fibers. Furthermore, many valleys and cracks were observed after simulated aging for 3,000 years, indicating severe degradation and damage of the commercial unprocessed Xuan paper. Compared with commercial unprocessed Xuan paper, commercial processed Xuan paper showed more ravines on cellulose fibers after simulated aging for 2,000 years. The presence of aluminum sulfate accelerated the acid hydrolysis of cellulose fibers. After simulated aging for 3,000 years, commercial processed Xuan paper was further degraded and damaged. The plant cellulose fibers fragmented into a large number of particles, and its tensile strength was further decreased. In comparison, the as-prepared inorganic fire-resistant analogous Xuan paper was of 100% inorganic origin with a high stability, and SEM images showed no obvious change of ultralong hydroxyapatite nanowires in the paper before aging and after a long period of simulated aging for 2,000 and 3,000 years, leading to the high retention rate of tensile strength (95.2% and 81.3%) after simulated aging for 2,000 and 3,000 years, respectively. In contrast, the average tensile strength retention rates of commercial unprocessed Xuan paper and processed Xuan paper were only 54.9% and 69.1%, respectively, after simulated aging for 2,000 years, and 40.4% and 55.8%, respectively, after simulated aging for 3,000 years.

The stability of whiteness is another important factor for the durability of Xuan paper. Figure 5.5 shows the whiteness stability of the inorganic fire-resistant analogous Xuan paper measured by the simulated accelerated heat aging method for up to 3,000 years, and the commercial Xuan paper sheets were used as control samples. The whiteness of the inorganic fire-resistant analogous Xuan paper was very high (92%), much higher than that of traditional Xuan paper, as shown in Figure 5.5 (70.5% for the unprocessed Xuan paper and 70.1% for processed Xuan paper). Although the fine handmade processes of traditional Xuan paper leads to a relatively high stability of whiteness compared with other kinds of traditional paper based on plant fibers, the whiteness of traditional Xuan paper still significantly reduced gradually during the simulated aging process. Figure 5.5a and b shows the decrease in whiteness of traditional Xuan paper with increasing simulated aging time. The whiteness decreased from an initial 70.5% to 47.3% and to 42.2% with a whiteness retention rate of 67.1% and 59.9% for commercial unprocessed Xuan paper after simulated aging for 2,000 and 3,000 years, respectively, and from an initial 70.1% to 46.4% and to 40.6% with a whiteness retention rate of 66.2% and 57.9% for commercial processed Xuan paper after simulated aging for 2,000 and 3,000 years, respectively. In contrast, the whiteness of the inorganic fire-resistant analogous Xuan paper exhibited a slight decrease from the initial 92% to 91.6% and to 86.7% with a whiteness retention rate as high as 99.6% and 94.2% even after simulated aging for 2,000 and 3,000 years, respectively (Figure 5.5c). The whiteness (86.7%) of the inorganic fire-resistant analogous Xuan paper after simulated aging for 3,000 years was still

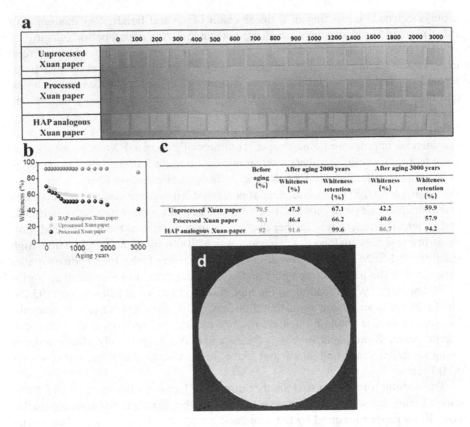

FIGURE 5.5 Evaluation of the whiteness stability of the inorganic fire-resistant analogous Xuan paper by the simulated accelerated heat aging method for up to 3,000 years compared with commercial Xuan paper. (a) Digital images of the three kinds of paper sheets with increasing simulated aging time up to 3,000 years; (b) whiteness changes of three kinds of paper sheets with increasing simulated aging time up to 3,000 years; (c) whiteness values and whiteness retention rates of three kinds of paper sheets after the simulated aging for 2,000 years and 3,000 years; (d) a digital image of an inorganic fire-resistant analogous Xuan paper sheet consisting of ultralong hydroxyapatite nanowires with a more clean surface after simulated accelerated heat aging for more than 10,000 years. (Reprinted with permission from reference [72])

much higher than that of traditional Xuan paper without aging (~70%), indicating the excellent whiteness stability of the inorganic fire-resistant analogous Xuan paper even after the simulated aging for up to 3,000 years.[72]

Further experiments showed that the whiteness of the inorganic fire-resistant analogous Xuan paper was still as high as 80.1% with a whiteness retention rate of 87.1% even after simulated aging for 10,000 years. In fact, the whiteness stability of the as-prepared inorganic fire-resistant analogous Xuan paper is related with the surface property of ultralong hydroxyapatite nanowires, that is, the amount of adsorbed oleate or oleic acid on the surface of ultralong hydroxyapatite nanowires.

The author's research group did the experiment using an inorganic fire-resistant analogous Xuan paper sheet prepared with ultralong hydroxyapatite nanowires with a more clean surface by the simulated accelerated heat aging method for more than 10,000 years, as shown in Figure 5.5d, and the inorganic fire-resistant analogous Xuan paper sheet prepared with ultralong hydroxyapatite nanowires with a more clean surface exhibited a high whiteness of 91.2% after the simulated accelerated heat aging for more than 10,000 years. In contrast, commercial unprocessed Xuan paper showed a very low whiteness of 26.3% after the simulated accelerated heat aging for more than 10,000 years.

The ink stability on Xuan paper is another important factor for the long-term preservation of precious paintings and calligraphy works. Figure 5.6 shows the ink stability on three kinds of paper sheets after simulated aging for 2,000 and 3,000 years. The experiments showed that the ink was still black, stable, and clearly visible on the three kinds of paper sheets after simulated aging for 2,000 and 3,000 years. Interestingly, the ink on the inorganic fire-resistant analogous Xuan paper was more clear and shiny after simulated aging for 2,000 and 3,000 years. Furthermore, the ink had penetrated into commercial unprocessed Xuan paper, and the ink was clearly visible on the back side of the unprocessed Xuan paper (Figure 5.6, bottom row). However, no ink was visible on the back side of the inorganic fire-resistant analogous Xuan paper (Figure 5.6, bottom row). The experimental results demonstrated that the inorganic fire-resistant analogous Xuan paper possessed much superior properties compared with traditional Xuan paper even after simulated aging for up to 3,000 years. The inorganic fire-resistant analogous Xuan paper is promising for applications in calligraphy, paintings, fire-resistant books, documents, and archives, and can well safeguard these precious paper-based works for the long-term safe preservation without the fear of being destroyed by fire and high temperatures.[72]

5.1.3. High-Temperature-Resistant Label Paper

High-temperature-resistant label paper is usually used in special high-temperature environments, such as high-temperature areas in steel plants and nuclear power plants. High-temperature-resistant label paper is generally made of polyphenylene sulfide, aramid, polyimide, and other organic substances that can withstand certain high temperatures. In some cases, the surface of the paper made from plant fibers is coated with high-temperature-resistant coating to make the high-temperature-resistant label paper. Commercial high-temperature-resistant label paper products usually have disadvantages such as poor high-temperature resistance and unsatisfactory flame-retardant performance.

The fire-resistant paper made from ultralong hydroxyapatite nanowires has excellent resistance to both fire and high temperatures with excellent thermal stability, and can be used as the high-performance high-temperature-resistant label paper which can withstand even temperatures greater than 1,000°C. In addition to the fire-resistant inorganic paper based on ultralong hydroxyapatite nanowires, considering various application environments with different temperatures, the author's research group also developed a new kind of fire-retardant and high-temperature-resistant

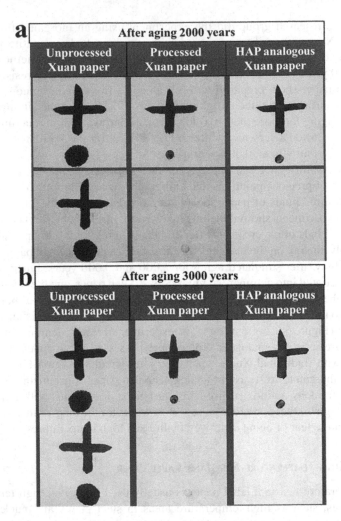

FIGURE 5.6 Ink stability on the inorganic fire-resistant analogous Xuan paper in comparison with the commercial traditional Xuan paper after simulated aging for 2,000 years (a) and 3,000 years (b). The top row shows the front surface of the paper, and the bottom row shows the back surface of the paper. (Reprinted with permission from reference [72])

label paper consisting of ultralong hydroxyapatite nanowires, aramid fibers and inorganic adhesive.[75] Aramid fiber paper is a special kind of paper made from aramid fibers with high-temperature resistance and electrical insulating property, and is widely used in electrical insulation, aviation, transportation, and many other fields. Aramid fibers are divided into meta-aramid fibers and para-aramid fibers, the main difference between the two types of aramid fibers being the different position where the amide group bonds with the C atom on the benzene ring. The thermal decomposition temperature of meta-aramid fibers is 370°C, and that of para-aramid fibers is 480°C.

Experimental results revealed that the addition of ultralong hydroxyapatite nanowires could significantly improve the properties of the high-temperature-resistant label paper. The high-temperature-resistant label paper exhibited excellent fire-retardant and high-temperature-resistant properties. Compared with the commercial high-temperature-resistant label paper, which was carbonized heavily at 400°C, the as-prepared high-temperature-resistant label paper is more suitable for applications in various high-temperature environments such as nuclear power plants, and iron and steel works.

The calendering process can effectively improve the mechanical strength, smoothness, and glossiness of the paper, which are important for the printing paper. The calendering process was adopted for the preparation of the high-temperature-resistant label paper. The basic physical properties of the as-prepared high-temperature-resistant label paper are shown in Figure 5.7. The experiments showed that the addition of ultralong hydroxyapatite nanowires with suitable amounts could enhance the tensile strength of the high-temperature-resistant label paper. When the weight ratio of hydroxyapatite/(hydroxyapatite + aramid) was 60%, the tensile strength of the high-temperature-resistant label paper reached the maximum value of 11.5 MPa (Figure 5.7a). For the preparation of the traditional cellulose fiber paper, the paper surface sizing and coating processes are adopted to realize high smoothness and high glossiness. Encouragingly, the high smoothness and high glossiness of the high-temperature-resistant label paper could be achieved only by calendering without the traditional sizing and coating processes. After calendering, the smoothness and glossiness of the pure aramid fiber paper without ultralong hydroxyapatite nanowires were 62.2 s and 5.3%, respectively. In contrast, the smoothness and glossiness of the high-temperature-resistant label paper could be significantly increased by increasing the content of ultralong hydroxyapatite nanowires (Figure 5.7b and c). The smoothness and glossiness of the high-temperature-resistant label paper with a hydroxyapatite/aramid weight ratio of 20/80 were 278.9 s and 12.7%, respectively, which were ~4.5 times and ~2.4 times those of the pure aramid fiber paper, respectively. The smoothness and glossiness of the high-temperature-resistant label paper with a hydroxyapatite/aramid weight ratio of 80/20 were 904 s and 24.1%, respectively, which were ~14.5 times and ~4.6 times those of the pure aramid fiber paper, respectively. Traditional cellulose fiber paper without the surface sizing and coating processes cannot achieve such high smoothness and glossiness.

The flame-retardant and high-temperature-resistant properties of the high-temperature-resistant label paper were investigated by the oxygen index and cone calorimetry. Oxygen index is defined as the minimum oxygen concentration required for flame combustion in a mixture gas of oxygen and nitrogen, which is the important parameter for the flame-retardant property of materials. The oxygen indexes of the high-temperature-resistant label paper sheets with different hydroxyapatite/aramid weight ratios are shown in Figure 5.8. The oxygen index of the pure aramid fiber paper without ultralong hydroxyapatite nanowires was measured to be 27.6%; therefore, the pure aramid fiber paper could not burn when it was away from the flame. The oxygen index of the high-temperature-resistant label paper was 32.2%, 34.9%, 43.4%, and 76.8% when the hydroxyapatite/aramid weight ratio was 20/80, 40/60,

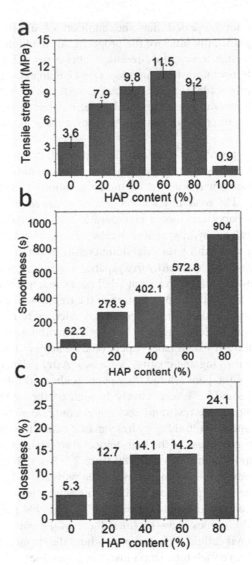

FIGURE 5.7 Physical properties of the high-temperature-resistant label paper sheets with different weight ratios of ultralong hydroxyapatite nanowires. (a) Tensile strength; (b) smoothness; (c) glossiness. (Reprinted with permission from reference [75])

60/40, and 80/20, respectively. The oxygen index of the high-temperature-resistant label paper significantly increased with increasing hydroxyapatite/aramid weight ratio, demonstrating the excellent flame-retardant property of the as-prepared high-temperature-resistant label paper.

The digital images of the high-temperature-resistant label paper sheets with different hydroxyapatite/aramid weight ratios before and after the cone calorimetry tests are shown in Figure 5.9a. The pure aramid fiber paper without ultralong hydroxyapatite

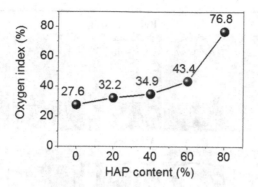

FIGURE 5.8 Oxygen index of the high-temperature-resistant label paper versus weight ratio of ultralong hydroxyapatite nanowires. (Reprinted with permission from reference [75])

nanowires was burned to ashes after the cone calorimetry test. However, the high-temperature-resistant label paper sheets with hydroxyapatite/aramid weight ratios ranging from 20/80 to 80/20 were intact and flat, indicating that the addition of ultralong hydroxyapatite nanowires could greatly enhance the high-temperature resistance of the high-temperature-resistant label paper. The heat release rate of materials is the most important parameter for predicting the combustion hazard and provides information on the propagation speed and burning degree of combustion. Figure 5.9b shows the heat release rate of the high-temperature-resistant label paper sheets with different hydroxyapatite/aramid weight ratios. The maximum heat release rate of the pure aramid fiber paper without ultralong hydroxyapatite nanowires was 193.9 kW m^{-2}, while the maximum heat release rate of the high-temperature-resistant label paper with a hydroxyapatite/aramid weight ratio of 20/80, 40/60, 60/40, and 80/20 was 119.6, 109.6, 82, and 65 kW m^{-2}, respectively, which obviously decreased with increasing hydroxyapatite/aramid weight ratio from 20/80 to 80/20. Similarly, the total heat release of the high-temperature-resistant label paper with a hydroxyapatite/aramid weight ratio of 0, 20%, 40%, 60%, and 80% was 4.91, 3.14, 2.56, 1.41, and 0.98 MJ m^{-2}, respectively, implying that the combustion intensity of the high-temperature-resistant label paper reduced with increasing hydroxyapatite/aramid weight ratio (Figure 5.9c). Figure 5.9d and e shows that the smoke produce rate and total smoke release of the high-temperature-resistant label paper greatly decreased with increasing hydroxyapatite/aramid weight ratio from 20/80 to 80/20. The total smoke release of the pure aramid fiber paper without ultralong hydroxyapatite nanowires was 84.2 m^2 m^{-2}; the total smoke release of the high-temperature-resistant label paper with a hydroxyapatite/aramid weight ratio of 20/80 was reduced to 52.9 m^2 m^{-2}; the total smoke release of the high-temperature-resistant label paper was as low as 16.6 m^2 m^{-2} when the hydroxyapatite/aramid weight ratio was 80/20, indicating that the addition of ultralong hydroxyapatite nanowires could significantly lower the amount of smoke produced during the burning process in fire.[75]

In order to further evaluate the fire safety of the high-temperature-resistant label paper, the fire performance index and fire growth index were analyzed. The fire

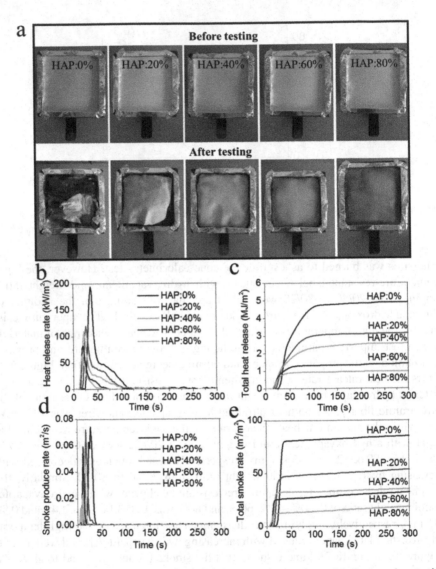

FIGURE 5.9 Thermal parameters of the high-temperature-resistant label paper sheets with different weight ratios of ultralong hydroxyapatite nanowires measured by cone calorimetry tests. (a) Digital images of the high-temperature-resistant label paper sheets before and after the cone calorimetry testing; (b) heat release rate; (c) total heat release; (d) smoke produce rate; (e) total smoke release. (Reprinted with permission from reference [75])

performance index is the ratio of time to ignition and maximum heat release rate, which is related to the flashover time. The higher the fire performance index, the longer the flashover time, which means there is a longer time for people to escape. Therefore, the high fire performance index is necessary in fire disaster control. The fire growth index is defined as the ratio of the maximum heat release rate and the time to reach the maximum heat release rate. The greater the fire growth index, the shorter the time to reach the maximum heat release rate, and the higher risk of the materials. The fire performance index of the high-temperature-resistant label paper increased with increasing hydroxyapatite/aramid weight ratio from 0 to 80/20, but the fire growth index exhibited an opposite trend, indicating that the higher the hydroxyapatite/aramid weight ratio, the lower the danger of the high-temperature-resistant label paper in fire.

Indicating signboards are necessary in high-temperature environments such as in nuclear power plants, and iron and steel works. The thermal stability performance of the as-prepared high-temperature-resistant label paper was compared with that of commercial high-temperature-resistant label paper, as shown in Figure 5.10. During the heating process, the whiteness of the as-prepared high-temperature-resistant label paper gradually decreased with increasing heating temperature from 200°C to 500°C for 30 minutes, but the dimension of the label paper had no obvious change. However, the white color of the commercial high-temperature-resistant label paper significantly deteriorated at 300°C; it was totally carbonized and turned black on heating at 400°C for 30 minutes and burned to ashes on burning at 500°C for 30 minutes (Figure 5.10). The experimental results indicated that the as-prepared

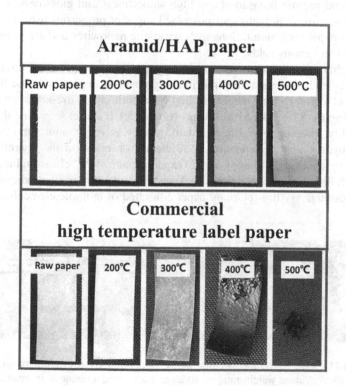

FIGURE 5.10 Thermal stability tests at various temperatures for 30 minutes of the as-prepared high-temperature-resistant label paper in comparison with a commercial high-temperature-resistant label paper. (Reprinted with permission from reference [75])

high-temperature-resistant label paper had excellent thermal stability and promising applications in various high-temperature environments.[75]

5.1.4. HIGHLY SMOOTH AND HIGHLY GLOSSY FIRE-RETARDANT PAPER

As discussed above, ultralong hydroxyapatite nanowires are the ideal building material for the new kind of fire-resistant paper, and a lot of research findings have been reported by the author's research group. However, properties such as smoothness and glossiness of the fire-resistant paper based on ultralong hydroxyapatite nanowires need to be further improved.

The author's research group developed a new kind of fire-retardant paper with good mechanical properties, excellent flame retardancy, ultrahigh smoothness, and high glossiness based on ultralong hydroxyapatite nanowires, cellulose fibers, and inorganic adhesive.[76] It was found that the addition of ultralong hydroxyapatite nanowires could significantly enhance the fire-retardant performance, smoothness, and glossiness of the cellulose fiber paper. Excitingly, the as-prepared fire-retardant paper exhibited ultrahigh smoothness and high glossiness without surface sizing and coating processes, while the surface sizing and coating processes are necessary for traditional papermaking to obtain high smoothness and glossiness. The highly smooth/glossy fire-retardant paper possessed superior properties over the pure cellulose fiber paper without ultralong hydroxyapatite nanowires and are promising for applications in various fields.

The highly smooth/glossy fire-retardant paper was composed of three components: ultralong hydroxyapatite nanowires, high beating degree cellulose fibers, and inorganic adhesive. Digital images of a highly smooth/glossy fire-retardant paper are shown in Figure 5.11a. The SEM images (Figure 5.11b and c) show that the surface of the highly smooth/glossy fire-retardant paper was very smooth and dense, and ultralong hydroxyapatite nanowires were clearly observed in the magnified SEM image. In many cases, ultralong hydroxyapatite nanowires self-assembled to form thicker bundles. The energy-dispersive X-ray spectroscopy results indicate that the highly smooth/glossy fire-retardant paper consisted of multiple elements, including

FIGURE 5.11 Characterization of the highly smooth/glossy fire-retardant paper with a hydroxyapatite/cellulose weight ratio of 80/20 and a certain amount of inorganic adhesive. (a) Digital images of the highly smooth/glossy fire-retardant paper; (b, c) SEM images of the surface of the highly smooth/glossy fire-retardant paper. (Reprinted with permission from reference [76])

Ca, P, O, C, Si, Al, S, and K. Ca, P, and O originated from ultralong hydroxyapatite nanowires; C and O were from cellulose fibers; and Si, Al, S, and K originated from the inorganic adhesive.

The basic physical properties of the highly smooth/glossy fire-retardant paper are shown in Figure 5.12. For traditional papermaking, a certain amount of inorganic fillers, such as talc powder and calcium carbonate, are usually added to the

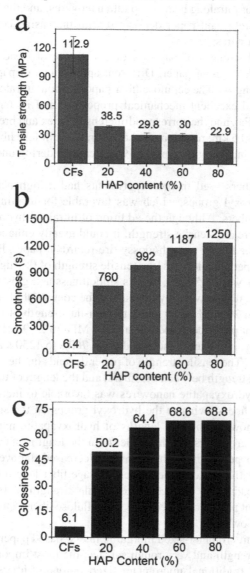

FIGURE 5.12 Physical properties of the highly smooth/glossy fire-retardant paper sheets with different hydroxyapatite/(cellulose + hydroxyapatite) weight ratios. (a) Tensile strength; (b) smoothness; (c) glossiness. (Reprinted with permission from reference [76])

paper pulp to improve the optical properties and reduce the cost. Fillers are a kind of micrometer-sized particles; the dosage of the fillers is generally about 20 wt.%, and excessive addition of fillers will affect the mechanical strength of the paper. Ultralong hydroxyapatite nanowires had diameters of about 10 nm and lengths of several hundred micrometers. The experimental results showed that the basis weight and tightness of the highly smooth/glossy fire-retardant paper increased with increasing content of ultralong hydroxyapatite nanowires, and the bulk of the highly smooth/glossy fire-retardant paper decreased with increasing content of ultralong hydroxyapatite nanowires.

The mechanical strength of paper is one of the most basic properties and is important for applications of paper. Different applications of paper require different mechanical strengths. The cellulose fiber paper without ultralong hydroxyapatite nanowires exhibited excellent mechanical properties, and its tensile strength was very high (112.9 MPa), and the corresponding tensile index and breaking length were 103.4 m N g^{-1} and 10.6 km, respectively; however, the cellulose fiber paper had a low smoothness (6.4 s) and low glossiness (6.1%) after calendering under a line pressure of 140 N mm^{-1}.

The cellulose fibers used in the experiments had a high beating degree with more exposed hydroxyl groups, which was favorable for forming hydrogen bonds between cellulose fibers. Although the addition of ultralong hydroxyapatite nanowires resulted in decreased tensile strength, it could greatly enhance the smoothness and glossiness of the highly smooth/glossy fire-retardant paper. By adding 20 wt.% ultralong hydroxyapatite nanowires, the tensile strength of the highly smooth/glossy fire-retardant paper was 38.5 MPa, and its smoothness and glossiness were as high as 760 s and 50.2%, respectively. By increasing the content of ultralong hydroxyapatite nanowires from 20 wt.% to 80 wt.%, the tensile strength of the highly smooth/glossy fire-retardant paper decreased from 38.5 MPa to 22.9 MPa, and its smoothness and glossiness significantly increased from 760 s to 1250 s and from 50.2% to 69.8%, respectively. The tensile strength of paper depends on the strength of the fiber itself, the bonding strength between the fibers, and the length of the fibers. The addition of ultralong hydroxyapatite nanowires was favorable to increasing the binding force between the fibers owing to the hydroxyl groups on the surface of ultralong hydroxyapatite nanowires, but the lengths of hydroxyapatite nanowires were only several hundred micrometers, much smaller than the lengths of cellulose fibers (2~3 mm) used in the papermaking, which greatly decreased the average length of the fibers, and the negative effect of decreased average fiber length on tensile strength was dominated, leading to the decreased tensile strength of the highly smooth/glossy fire-retardant paper compared with the cellulose fiber paper without ultralong hydroxyapatite nanowires.

The traditional micrometer-sized inorganic fibers–based paper usually has a very low mechanical strength and very rough surface; the most important reason is that the surface of the traditional micrometer-sized inorganic fibers does not contain bonding groups such as hydroxyl groups. Compared with traditional inorganic fibers, ultralong hydroxyapatite nanowires had a large number of hydroxyl groups on the surface, which was the reason why the highly smooth/glossy fire-retardant paper still exhibited good mechanical strength when the content of ultralong hydroxyapatite nanowires was as high as 80 wt.%. In order to improve the smoothness and glossiness, surface sizing, coating, and calendering are needed for traditional cellulose fiber paper. In contrast, only calendering process without surface sizing and coating could achieve ultrahigh smoothness and high glossiness for a highly smooth/glossy fire-retardant paper, which is related to the nanoscale ultrafine structure of ultralong hydroxyapatite nanowires.

Figure 5.13a shows the appearance of the highly smooth/glossy fire-retardant paper sheets with different contents of ultralong hydroxyapatite nanowires after the burning test. The cellulose fiber paper without ultralong hydroxyapatite nanowires was completely burned to ashes during the burning test. However, the highly smooth/glossy fire-retardant paper sheets with different weight ratios of ultralong hydroxyapatite nanowires exhibited excellent fire-retardant property, and the carbonization height of the paper became shorter with increasing content of ultralong hydroxyapatite nanowires, indicating that the flame-retardant performance of the paper could be improved by adding ultralong hydroxyapatite nanowires.

The oxygen index values of the highly smooth/glossy fire-retardant paper sheets with different weight ratios of ultralong hydroxyapatite nanowires are

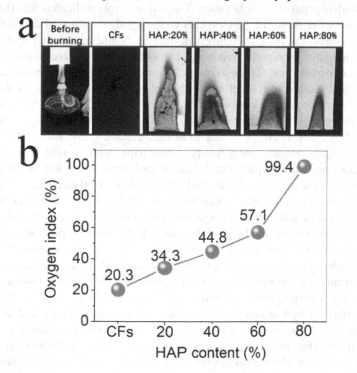

FIGURE 5.13 (a) Digital images of different paper sheets after the burning test; (b) oxygen index values of the highly smooth/glossy fire-retardant paper sheets with different hydroxyapatite/(cellulose + hydroxyapatite) weight ratios. (Reprinted with permission from reference [76])

shown in Figure 5.13b. The oxygen index of the cellulose fiber paper without ultralong hydroxyapatite nanowires was 20.3%, indicating that the cellulose fiber paper was flammable in air. In contrast, the oxygen index of the highly smooth/glossy fire-retardant paper sheets was 34.3%, 44.8%, 57.1%, and 99.4% when the hydroxyapatite/(cellulose + hydroxyapatite) weight ratio was 20%, 40%, 60%, and 80%, respectively, indicating the flame retardancy of the paper could be greatly enhanced with increasing the weight ratio of ultralong hydroxyapatite nanowires. The experiments showed that the highly smooth/glossy fire-retardant paper could quickly extinguish the burning fire when it was taken away from the flame. The flame-retardant mechanism of the highly smooth/glossy fire-retardant paper is related to the formation of the nonflammable inorganic dense layer of ultralong hydroxyapatite nanowires and inorganic adhesive after burning. The nonflammable inorganic dense layer could inhibit the gas diffusion (oxygen and the combustible gas from the decomposition of cellulose). Furthermore, the non-flammable inorganic dense layer could inhibit the transfer of the flame heat to the inner cellulose fibers.[76]

Cone calorimetry test was performed to evaluate the fire-retardant properties of the highly smooth/glossy fire-retardant paper with different hydroxyapatite/(cellulose + hydroxyapatite) weight ratios. The experiments indicated that the highly smooth/glossy fire-retardant paper sheets were slightly carbonized due to the presence of cellulose fibers. The highly smooth/glossy fire-retardant paper with 20 wt.% ultralong hydroxyapatite nanowires showed an obvious size contraction after the cone calorimetry test. The maximum heat release rate, mean heat release rate, and total heat release of the cellulose fiber paper without ultralong hydroxyapatite nanowires were 250.4 kW m^{-2}, 12.0 kW m^{-2}, and 3.59 MJ m^{-2}, respectively. The maximum heat release rate (117.7, 69.4, 51.5, 29.6 kW m^{-2}), mean heat release rate (5.6, 5.3, 2.7, 2.1 kW m^{-2}), and total heat release (1.87, 1.61, 0.97, 0.70 MJ m^{-2}) of the highly smooth/glossy fire-retardant paper were significantly decreased with increasing hydroxyapatite/(cellulose + hydroxyapatite) weight ratio (20%, 40%, 60%, 80%, respectively). The total smoke release of the cellulose fiber paper without ultralong hydroxyapatite nanowires was 8.44 m^2 m^{-2}. However, adding a suitable amount of ultralong hydroxyapatite nanowires in the paper could greatly decrease the amount of smoke produced during the burning process in fire. The total smoke release of the highly smooth/glossy fire-retardant paper with 20 wt.% ultralong hydroxyapatite nanowires was very small (0.35 m^2 m^{-2}). When the weight ratio of ultralong hydroxyapatite nanowire was 80 wt.%, the total smoke release of the highly smooth/glossy fire-retardant paper was 14.76 m^2 m^{-2}. This result can be explained by the fact that excessive addition of ultralong hydroxyapatite nanowires would lead to the inadequate combustion of cellulose fibers and higher smoke output. The fire growth index value of the cellulose fiber paper without ultralong hydroxyapatite nanowires was 16.7. However, the fire growth index of the highly smooth/glossy fire-retardant paper decreased with increasing weight ratio of ultralong hydroxyapatite nanowires from 20 wt.% to 80 wt.%, indicating that a higher weight ratio of ultralong hydroxyapatite nanowires could improve the safety of the highly smooth/glossy fire-retardant paper.[76]

5.2. APPLICATIONS OF FIRE RESISTANCE AND HEAT INSULATION

The fire-resistant paper based on ultralong hydroxyapatite nanowires has excellent resistance performance to both fire and high temperatures, and has low thermal conductivity and excellent heat insulation properties. Therefore, the fire-resistant paper can be used in various high-temperature environments, such as fire-resistant paper tape for electric cables and fiber-optic cables, automatic fire alarm fire-resistant wallpaper, and fire-resistant doors.

5.2.1. FIRE-RESISTANT PAPER TAPE FOR ELECTRIC CABLES AND FIBER-OPTIC CABLES

If the "Internet" is called as the "information highway" of today's human society, then electric cables and fiber-optic cables are the cornerstone of the "information highway." With the rapid development of the communication technology, electric cables and fiber-optic cables are widely used in various fields of our daily life and work. In the event of a fire, keeping the communication at the fire disaster site smooth is essential to reduce casualties and property losses, and this emphasizes the importance of the fire-resistant electric cables and fiber-optic cables. Fire-resistant electric cables and fiber-optic cables are mainly composed of a cable core (optical fiber or metal wire), flame-resistant tape, reinforcing parts, and outer sheath. In the event of a fire disaster, the protection of the cable core and maintenance of smooth communication is mainly achieved by the fire-resistant tape inside the electric cables and fiber-optic cables. Although different kinds of the commercial flame-retardant tapes are used in electric cables and fiber-optic cables, the fire resistance and heat insulation properties of these commercial flame-retardant tapes are usually not satisfactory, and the internal temperatures of the electric cables and fiber-optic cables are high in the flame, which may destroy the cables and affect the communication quality. In some cases, some commercial flame-retardant tapes also release toxic gases and smoke which endanger the lives of the people at the scene of the fire disaster.

The author's research group developed a fire-resistant paper tape made from network-structured ultralong hydroxyapatite nanowires, glass fibers, and inorganic adhesive for application in fireproof electric cables and fiber-optic cables.[58] The experiments demonstrated that the fire-resistant paper tape exhibited good mechanical properties (tensile strength ~16 MPa), high biocompatibility, environment friendliness, and high thermal stability at temperatures up to 1,000°C. In addition, the fire-resistant paper tape showed a low thermal conductivity and excellent heat insulation performance which were competitive to the conventional insulation materials.

The as-prepared fire-resistant paper tapes with different weight ratios of ultralong hydroxyapatite nanowires exhibited excellent fire-resistant performance and high thermal stability. Both components of ultralong hydroxyapatite nanowires and glass fibers were good thermal insulation materials with low thermal conductivities. The thermal conductivity of the ultralong hydroxyapatite nanowires and glass fibers was ~0.04 W m^{-1} K^{-1} and 0.11 W m^{-1} K^{-1}, respectively. The thermal conductivity of the fire-resistant paper tape with 80 wt.% ultralong hydroxyapatite nanowires was 0.10 W m^{-1} K^{-1} (25°C) and 0.14 W m^{-1} K^{-1} (600°C).

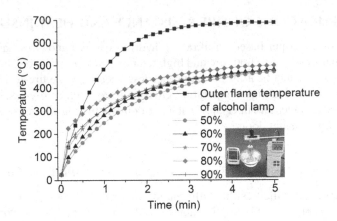

FIGURE 5.14 Thermal insulation properties of the fire-resistant paper tapes with weight ratios of ultralong hydroxyapatite nanowires from 50 wt.% to 90 wt.%. (Reprinted with permission from reference [58])

The thermal insulation properties of the fire-resistant paper tapes with different weight ratios of ultralong hydroxyapatite nanowires were investigated, and the experimental results are shown in Figure 5.14. Without the protection of the fire-resistant paper tape, the temperature increased rapidly and reached the highest temperature (~687°C) within 4 minutes by heating in the flame of an alcohol lamp. When the fire-resistant paper tapes with different weight ratios of ultralong hydroxyapatite nanowires (50–90 wt.%) were used as the protection layer, the temperature increased slowly. In the case of the fire-resistant paper tape with 50 wt.% ultralong hydroxyapatite nanowires, the temperature reached 470°C after heating for 5 minutes, which was 217°C lower than that without the fire-resistant paper tape as the protection layer. With increasing weight ratio of ultralong hydroxyapatite nanowires from 50 wt.% to 90 wt.%, the temperature had no obvious change and was close to 470°C after heating for 5 minutes. On the other hand, the paper thickness also had an effect on the heat insulation performance of the fire-resistant paper tape. The thicker the paper tape, the better the heat insulation performance. The thickness of the fire-resistant paper tape with a weight ratio of ultralong hydroxyapatite nanowires of 50 wt.%, 60 wt.%, 70 wt.%, 80 wt.%, 90 wt.% was 229 μm, 205 μm, 195 μm, 190 μm, 195 μm, respectively, and the paper densities were in the range of 0.43~0.50 g cm^{-3}.

As discussed above, the thermal conductivity of the fire-resistant paper tape was low; thus the fire-resistant paper tape is a promising fire-resistant heat insulation material. Superior to conventional heat insulation materials, such as aluminum silicate fiber paper and glass fiber paper, the as-prepared fire-resistant paper tape exhibited superior mechanical properties (tensile strength ~16 MPa), and it was smooth on the surface. In addition, the fire-resistant paper tape had excellent nonflammability, high thermal stability, and excellent heat insulation performance, which is promising for application as a fireproof and heat insulation protection layer for the fire-resistant electric cables and fiber-optic cables.

Compared with other kinds of protection materials for the fire-retardant electric cables and fiber-optic cables, the fire-resistant paper tape (for example, 80 wt.% ultralong hydroxyapatite nanowires, basis weight 200 g m^{-2}) showed much better fire-resistant and heat insulation properties, as shown in Figure 5.15a. Without the heat insulation protection layer, the temperature increased rapidly with increasing heating time, and the temperature was ~687°C after heating for 5 minutes. In comparison, using the fire-resistant paper tape as the protection layer, the temperature increased slowly and the temperature was 376°C after heating for 5 minutes, which is 311°C lower than that without using the heat insulation protection layer. Other kinds of the protection materials were also tested, such as the glass fiber paper,

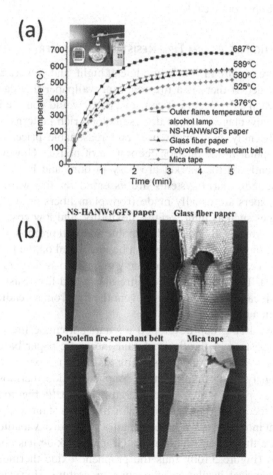

FIGURE 5.15 (a) Thermal insulation properties of the fire-resistant paper tape based on ultralong hydroxyapatite nanowires (NS-HANWs/GFs) and other kinds of traditional protection materials; (b) the appearances of the fire-resistant paper tape and other kinds of traditional protection materials after heating for 5 minutes in the flame of an alcohol lamp. (Reprinted with permission from reference [58])

polyolefin fire-retardant belt, and mica tape, which possessed similar thickness to that of the fire-resistant paper tape. The temperatures for the glass fiber paper, polyolefin fire-retardant belt, and mica tape were 589°C, 580°C, and 525°C, respectively, after heating for 5 minutes. In comparison, the temperature was as low as 376°C after heating for 5 minutes using the fire-resistant paper tape as the protection layer. The experiments showed that the fire-resistant paper tape exhibited much better fire resistance and heat insulation properties than those of the traditional protection materials. Moreover, the fire-resistant paper tape was intact after heating, but other traditional protection materials were destroyed to some extent (Figure 5.15b). Therefore, it is expected that the fire-resistant paper tape based on ultralong hydroxyapatite nanowires is promising for application as the protection layer for the fireproof electric cables and fiber-optic cables.[58]

5.2.2. AUTOMATIC FIRE ALARM FIRE-RESISTANT WALLPAPER

Just imagine if there is a fire in your home late at night, and you are asleep, then you are in extreme danger. If there is a fire-resistant wallpaper that can automatically trigger an alarm when a fire occurs, it can save lives, would you use it?

Wallpaper has advantages such as diverse colors, rich patterns, luxurious style, environmentally friendly, convenient use, and reasonable price. Nowadays, it is more and more popular in the interior decoration of houses. However, many kinds of wallpaper currently on the market are easy to burn and have safety concerns. On the other hand, fire alarm systems are essential for fire warning and rescue. Commercial wallpapers are usually made from plant fibers or synthetic polymers, and have advantages such as light weight, flexibility, and low cost. However, commercial wallpapers are usually highly flammable and will promote the spread of fire in a fire disaster. If the fire alarm system can be integrated on the fire-resistant wallpaper, the fire alarm can be sent out in a timely manner to avoid personnel casualties and property losses. Therefore, the smart fire alarm and fire-resistant wallpaper is desirable because it can simultaneously prevent the fire from spreading and send out immediate alerts in a fire disaster.

The author's research group developed a smart automatic fire alarm fire-resistant wallpaper consisting of the fire-resistant inorganic paper based on ultralong hydroxyapatite nanowires that could resist fire and high temperatures and maintain its structural integrity in the flame, and a graphene oxide thermosensitive sensor that could rapidly respond to the high temperature of fire.[77]. Figure 5.16a shows the design scheme of the smart automatic fire alarm fire-resistant wallpaper consisting of the fire-resistant inorganic paper based on ultralong hydroxyapatite nanowires and the graphene oxide thermosensitive sensor. Graphene oxide has abundant oxygen-containing groups (Figure 5.16b); thus the graphene oxide thermosensitive sensor is in a state of electrical insulation at room temperature. However, the graphene oxide thermosensitive sensor becomes electrically conductive at high temperatures because the oxygen-containing groups of graphene oxide could be rapidly removed at high temperatures, resulting in the transformation of graphene oxide from an electrically insulated state into a highly conductive one. In this way, the alarm lamp and

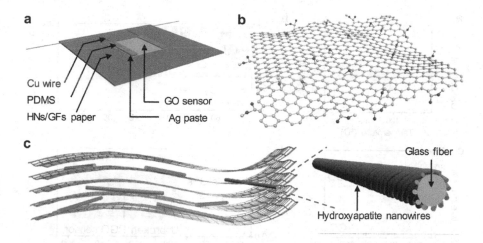

FIGURE 5.16 (a) Schematic illustration of the smart automatic fire alarm fire-resistant wallpaper; (b) graphene oxide (GO) with abundant oxygen-containing groups; (c) the smart automatic fire alarm fire-resistant wallpaper exhibited a multilayered structure. (Reprinted with permission from reference [77])

buzzer connected with the graphene oxide thermosensitive sensor could send out the alerts to people immediately for taking emergency actions. In the smart automatic fire alarm fire-resistant wallpaper, the glass fibers were wrapped with ultralong hydroxyapatite nanowires, and the fire-resistant inorganic paper had a multilayered structure (Figure 5.16c).

For the preparation of the graphene oxide thermosensitive sensor, graphene oxide was well dispersed in deionized water by vigorous stirring and ultrasonication, and the stable graphene oxide aqueous ink was obtained. The graphene oxide thermo-sensitive sensor was installed on the fire-resistant inorganic paper by a facile drop-casting of the graphene oxide aqueous ink. The copper wires were connected to the two edges of the thermosensitive sensor as external electrodes using silver paste and poly(dimethylsiloxane).

The sensitivity of the as-prepared polydopamine-modified graphene oxide ther-mosensitive sensor was studied. The polydopamine-modified graphene oxide ther-mosensitive sensor was weakly electrically conductive ($\sigma = 11.19$ S m^{-1}) owing to a certain degree of reduction of graphene oxide by polydopamine. However, the alarm lamp connected with the thermosensitive sensor at a low applied voltage (2.5 V) showed only weak visible light, and the alarm buzzer could not send out any audible sound. The bright light and loud sound were sent out from the polydopamine-mod-ified graphene oxide thermosensitive sensor after thermal treatment at 150°C for 3 minutes, and the electrical conductivity was 188.83 S m^{-1}. The energy-dispersive X-ray spectroscopy analysis revealed the increased atomic percentage of carbon and the decrease of atomic percentage of oxygen with increasing thermal treatment tem-perature, as shown in Figure 5.17a. The accurate thermal responsive temperature of the polydopamine-modified graphene oxide thermosensitive sensor was determined

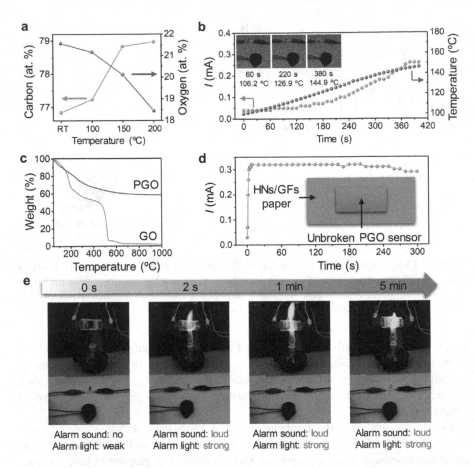

FIGURE 5.17 (a) Atomic percentages of carbon and oxygen of the polydopamine (PGO)-modified graphene oxide thermosensitive sensor after thermally treating at different temperatures; (b) electrical current (I) through the polydopamine-modified graphene oxide thermosensitive sensor with increasing temperature (the inset shows digital images of an alarm lamp and alarm buzzer connected with the polydopamine-modified graphene oxide thermosensitive sensor at different temperatures); (c) thermogravimetric curves of the graphene oxide and polydopamine-modified graphene oxide thermosensitive sensors; (d) electrical current through the polydopamine-modified graphene oxide thermosensitive sensor on the fire-resistant paper in the alcohol flame with time (the inset shows the intact polydopamine-modified graphene oxide thermosensitive sensor on the fire-resistant paper after the burning test in the alcohol flame); (e) real-time monitoring of the alarm lamp, alarm buzzer, and polydopamine-modified graphene oxide thermosensitive sensor on the fire-resistant paper during the testing process in the alcohol flame. (Reprinted with permission from reference [77])

by real-time monitoring of the electrical current change with increasing temperature in an electric oven (Figure 5.17b). The uptrend of electrical current through the polydopamine-modified graphene oxide thermosensitive sensor was similar to the graphene oxide thermosensitive sensor. Compared with the faster change of electrical current through the graphene oxide thermosensitive sensor with increasing temperature, the gentle increase of electrical current through the polydopamine-modified graphene oxide thermosensitive sensor resulted from its enhanced thermal stability. The thermal responsive temperature (electrical current ~0.10 mA) of the polydopamine-modified graphene oxide thermosensitive sensor was 126.9°C, which was much lower than that of the graphene oxide thermosensitive sensor (231.3°C). The thermogravimetric curve showed that the residue weight of the polydopamine-modified graphene oxide thermosensitive sensor was 58.69% of the initial weight at 1,000°C in air, and the residue weight of the graphene oxide thermosensitive sensor was only 1.96% of the initial weight, indicating the enhanced thermal stability of the polydopamine-modified graphene oxide thermosensitive sensor compared with the graphene oxide thermosensitive sensor (Figure 5.17c). Moreover, the polydopamine-modified graphene oxide thermosensitive sensor could rapidly respond to the high temperature of fire. The thermal response time to fire was very short (only several seconds). In addition, the polydopamine-modified graphene oxide thermosensitive sensor could steadily work and send out an alarm in the flame for a relatively long period of time (at least 5 minutes, Figure 5.17d and e). The electrical current through the polydopamine-modified graphene oxide thermosensitive sensor reached a maximal value of 0.32 mA at 8 seconds, and then it was relatively stable in the flame (more than 5 minutes). The experiments showed that the sensitivity and thermal stability of the thermosensitive sensor could be significantly enhanced by modifying graphene oxide with polydopamine. The polydopamine-modified graphene oxide thermosensitive sensor exhibited a lower thermal responsive temperature, faster response, and longer alarm time compared with the graphene oxide thermosensitive sensor without surface modification with polydopamine.[77]

For the interior decoration of houses, it is highly desirable that the automatic fire alarm fire-resistant wallpaper can be processed into various colored patterns and shapes. The fire-resistant wallpaper with a white color, mechanical robustness, high flexibility, and excellent fire resistance can satisfy these requirements. As shown in Figure 5.18a and b, the fire-resistant wallpaper could be folded into complex shapes such as the paper airplane and paper crane. By adding a small amount of dyes in the initial aqueous suspension for papermaking, the automatic fire alarm fire-resistant wallpaper with different colors like purple, pink, red, blue, yellow, and orange could be obtained (Figure 5.18c). In addition, the fire-resistant wallpaper could be printed in various colors using a commercial printer, and the colorful patterns and images could be clearly printed on the fire-resistant wallpaper (Figure 5.18d–f).

5.3. MULTIMODE ANTI-COUNTERFEITING

Nowadays, the increasing prevalence of fake and shoddy goods has severely affected people's daily work and life, and even threatened people's health and safety. Prominent feature of these counterfeit and shoddy products is the use of the same or similar patterns and styles in the brand names, trademark, and packaging. What is more severe is the forgery of currency and important documents. However, many available anti-counterfeiting technologies cannot effectively prevent the forgery. Anti-counterfeiting technologies have increasingly demonstrated their importance

FIGURE 5.18 (a, b) The fire-resistant wallpaper could be folded into complex shapes such as a paper airplane (a) or paper crane (b); (c) the fire-resistant wallpaper dyed with different colors; (d–f) various colorful patterns and images were printed on the fire-resistant wallpaper using a commercial printer. (Reprinted with permission from reference [77])

in identifying and preventing counterfeit and shoddy products. The addition of anti-counterfeiting labels on products has become an effective measure for manufacturers to protect their products.

At present, the fluorescent anti-counterfeiting paper is one of the commonly used anti-counterfeiting materials. The fluorescent anti-counterfeiting paper has advantages such as convenient use and easy identification, and is widely used in banknotes, securities, certificates, and anti-counterfeiting packaging. The traditional fluorescent anti-counterfeiting paper is usually prepared by mixing plant fibers with the luminescent constituent or coating the luminescent material on the paper surface to realize the fluorescent anti-counterfeiting function. However, the traditional fluorescent anti-counterfeiting paper is composed of flammable plant fibers with poor thermal stability, which can be easily carbonized at high temperatures and burned in fire, and the paper will gradually turn yellow under the action of air and light. On the other hand, the counterfeiting methods are becoming more and more sophisticated, and it is difficult for a single fluorescent anti-counterfeiting technology to achieve effective and high-security anti-counterfeiting purposes, and any single or low-tech anti-counterfeiting method is difficult to effectively prevent counterfeiting. As a result, a high-security anti-counterfeiting technology that integrates multiple anti-counterfeiting methods is increasingly popular. The higher the technological content and level of the anti-counterfeiting technology, the greater the difficulty of counterfeiting and the better the anti-counterfeiting effect. Therefore, the development of a new high-security multimode anti-counterfeiting technology that integrates a variety of anti-counterfeiting methods has important significance and high application value.

As discussed in Section 4.6, the major weaknesses of traditional photoluminescent paper are its poor thermal stability and flammability. Traditional photoluminescent paper made from plant fibers can be easily ruined by fire. The author's research group adopted rare earth ion–doped ultralong hydroxyapatite nanowires modified with sodium oleate as the raw material for the preparation of the waterproof photoluminescent fire-resistant paper for the application in multimode anti-counterfeiting.[63] The experiments showed that the waterproof photoluminescent fire-resistant paper doped with rare earth ions exhibited a high thermal stability. As show in Figure 5.19a, the common commercial paper made from plant fibers was carbonized and the color

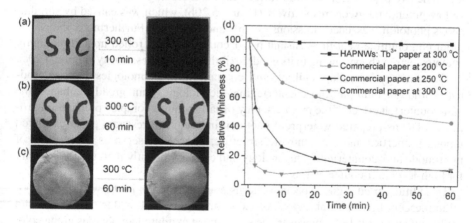

FIGURE 5.19 (a–c) Digital images of two kinds of paper sheets before and after thermal treatment: (a) a common commercial paper made of plant fibers; (b) a photoluminescent fire-resistant paper consisting of ultralong hydroxyapatite nanowires doped with 5 mol% Tb^{3+} ions, and it was writable; (c) the photoluminescent fire-resistant paper consisting of ultralong hydroxyapatite nanowires doped with 5 mol% Tb^{3+} ions exhibited green color under UV irradiation (~365 nm) before and after thermal treatment at 300°C for 1 hour; (d) whiteness change of a common commercial paper based on plant fibers, and a photoluminescent fire-resistant paper consisting of ultralong hydroxyapatite nanowires doped with 5 mol% Tb^{3+} ions after thermal treatment at different temperatures. (Reprinted with permission from reference [63])

changed from white to black after thermal treatment at 300°C for 10 minutes. However, the waterproof photoluminescent fire-resistant paper exhibited no obvious change and the text was still clearly visible after thermal treatment at 300°C for 1 hour (Figure 5.19b). The photoluminescence emission properties of the waterproof photoluminescent fire-resistant paper doped with Tb^{3+} ions before and after thermal treatment are shown in Figure 5.19c and d, indicating that there was no obvious change in the photoluminescence color (green) and intensity of the waterproof photoluminescent fire-resistant paper before and after thermal treatment. Figure 5.19d shows that the whiteness of the common commercial paper sheets decreased significantly after thermal treatment at 200°C and 250°C and dramatically decreased after thermal treatment at 300°C for only 5 minutes. In contrast, the whiteness of the waterproof photoluminescent fire-resistant paper had no obvious change even after thermal treatment at 300°C for 1 hour.

The waterproof photoluminescent fire-resistant paper showed a strong green color (doped with Tb^{3+} ions) or red color (doped with Eu^{3+} ions) under irradiation with a UV lamp (~365 nm). In addition, the waterproof photoluminescent fire-resistant paper sheets with tunable photoluminescence properties were prepared by adjusting the weight ratios of Eu^{3+}-doped ultralong hydroxyapatite nanowires to Tb^{3+}-doped ultralong hydroxyapatite nanowires, and five kinds of paper sheets with weight ratios of Eu^{3+}-doped ultralong hydroxyapatite nanowires to Tb^{3+}-doped ultralong hydroxyapatite nanowires ranging from 10:0 to 0:10 were prepared. As shown in Figure 5.20a, all of the paper sheets exhibited a white color under visible light with no obvious difference; however, they were totally different under UV irradiation (~365 nm). The five paper sheets showed various colors, including red, red-orange, orange, yellow-green, and green, respectively (Figure 5.20b), which was caused by simultaneous photoluminescence emissions of Tb^{3+} and Eu^{3+} ions. Furthermore, the waterproof photoluminescent fire-resistant paper could be easily rolled up and randomly cut into desired shapes owing to its good flexibility and processability (Figure 5.20c).

One strategy is to adopt multiple anti-counterfeiting technologies into one product. Although multiple anti-counterfeiting technologies can greatly enhance the anti-counterfeit effect of the products, they usually require complex preparation processes. The as-prepared waterproof photoluminescent fire-resistant paper exhibited unique properties and the nanowire structure, which can serve as a new kind of multimode anti-counterfeiting technology and can significantly increase the security for high-level anti-counterfeiting.

In addition, predesigned special patterns could be hidden in the waterproof photoluminescent fire-resistant paper. As shown in Figure 5.21, a star pattern was not visible under visible light; however, the star pattern exhibited an obvious green color and the other areas showed a red color under UV irradiation (~365 nm). This result indicated that the visibility of the predesigned pattern could be turned on and off using a single UV lamp, which can be used for the anti-counterfeiting purpose.

Moreover, the anti-counterfeit effect of the waterproof photoluminescent fire-resistant paper could greatly enhanced by its unique physical properties. For example, waterproofness and fire-resistance tests can be used to check the facticity. As shown in Figure 5.21, when the waterproof photoluminescent fire-resistant paper was immersed in water dyed with methylene blue and taken out from the water after

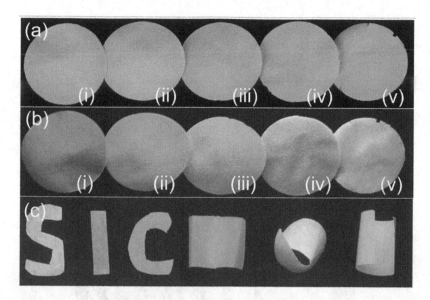

FIGURE 5.20 Digital images of the photoluminescent fire-resistant paper sheets doped with Eu^{3+} and Tb^{3+} ions with different weight ratios. (a) The paper sheets were all white under visible light; (b) the paper sheets exhibited different colors from red to green under UV irradiation (~365 nm) from (i) to (v) with the weight ratios of Eu^{3+}-doped ultralong hydroxyapatite nanowires to Tb^{3+}-doped ultralong hydroxyapatite nanowires were 10:0, 5:5, 3:7, 1:9, and 0:10, respectively; (c) the photoluminescent fire-resistant paper sheets with various shapes and different colors under UV irradiation (~365 nm). (Reprinted with permission from reference [63])

a few seconds, the paper did not get wet or colored and could maintain its white color. Furthermore, when the waterproof photoluminescent fire-resistant paper was exposed to fire, it could be well preserved without obvious damage owing to its excellent fire-resistance properties. Furthermore, the unique nanostructure of the waterproof photoluminescent fire-resistant paper could also be used in high-level security anti-counterfeiting. It is obvious that the nanowire structure of the waterproof photoluminescent fire-resistant paper was quite different from the microstructure of the common commercial paper made from plant fibers. The above experimental results demonstrated that the waterproof photoluminescent fire-resistant paper is a promising material for high-security advanced anti-counterfeiting applications.

The new waterproof photoluminescent fire-resistant paper integrated multiple functions. It is waterproof and environment friendly and has high-temperature resistance, fire resistance, and adjustable photoluminous color. It can be easily distinguished from ordinary flammable anti-counterfeiting labels as long as it is lit with a lighter. Its unique nanoscale microstructure can effectively distinguish it from ordinary anti-counterfeiting paper, and further improve its anti-counterfeiting effect and safety. To use this material in the anti-counterfeiting field, the waterproof photoluminescent fire-resistant paper sheets can be prepared into desired shapes, and they can be glued onto the products directly or onto the packages of the products.

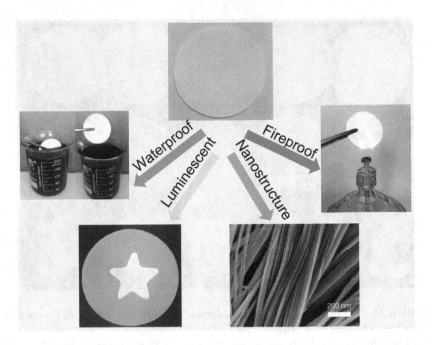

FIGURE 5.21 Illustration of the waterproof photoluminescent fire-resistant paper for multimode anti-counterfeiting application. (Reprinted with permission from reference [63])

Moreover, the waterproof photoluminescent fire-resistant paper can also be used as a writing paper and printing paper for anti-counterfeiting documents and certificates. The new multimode anti-counterfeiting fire-resistant paper has high-efficiency and high-security anti-counterfeiting effects, and is expected to be used in the anti-counterfeiting field to prevent forgery of banknotes, securities, certificates, etc., and also in anti-counterfeiting labels and anti-counterfeiting packaging.[63]

5.4. ENCRYPTION AND DECRYPTION FOR SECRET INFORMATION

In movies and TV dramas, you may sometimes see such storylines. In order to transmit secret information without being discovered by others, intelligence personnel use rice soup to write on paper. After the rice soup is dried, it looks like a blank sheet of paper, and the secret information on the paper is not visible at all. The personnel who receive the secret information soak the paper containing the secret information into iodine ethanol solution, and the secret information on the paper is revealed. The method of using rice soup to write on paper and then using iodine to display the text is a traditional encryption and decryption technology. The principle is that rice soup contains starch, and the starch will turn blue when it comes into contact with iodine due to the former's molecular structure.

The encryption and decryption technologies are of great significance in the fields of information protection, anti-counterfeiting, and data storage. In recent years, researchers have investigated the applications of photoluminescent materials, including organic dyes, quantum dots, and rare earth element–doped materials, in the fields of information protection, anti-counterfeiting, and security. Photoluminescent materials are usually added to solvents to prepare photoluminescent security inks that can be used for printing. However, photoluminescent security inks still have some shortcomings, for example: (1) the formulation parameters of the ink usually need to be optimized before use; (2) some inks containing organic dyes have poor stability; (3) the synthesis of some quantum dots and lanthanide-doped materials may introduce heavy metals, which may cause potential toxicity and environmental problems; and (4) although photoluminescent safety inks are invisible to human eyes under ambient light, it can be detected under ultraviolet light or near infrared light. Therefore, the security level of photoluminescent security inks needs to be further improved in order to be better used for the protection of secret information.

In recent years, stimulus-responsive materials that can cause color changes have attracted widespread attention. The stimuli used to trigger the color change include light, steam, mechanical force, pH value, temperature, electric field, and magnetic field, etc. However, many stimuli are not easily available in daily life, making it difficult to decrypt the encrypted information conveniently and quickly in practical applications. However, the rapid development of modern science and technology, especially nanotechnology, provides the possibility for the emergence of convenient and fast encryption and decryption technologies.

Recently, the author's research group developed a new type of secret paper using ultralong hydroxyapatite nanowires and plant fibers.[78] With regard to appearance, the as-prepared secret paper was as white as ordinary paper, and it exhibited excellent flexibility and processibility, as shown in Figure 5.22. The secret paper was investigated for application in fast encryption and fast decryption of secret information. White vinegar, a common inexpensive cooking material, was used as the invisible and safe ink to write on the new type of secret paper. After the secret paper containing the secret information written with white vinegar was dried, it was completely invisible to human eyes under natural light or lamplight, and it looked like just a piece of ordinary white paper. In this way, the encryption process of secret information was completed. The decryption process of the secret information on the new type of secret paper was also fast and simple. Once the secret paper was burned in fire for a few seconds, the secret information could be displayed and read, as shown in Figure 5.23. Upon exposure to fire, the plant fiber paper was rapidly ignited and incinerated in several seconds (Figure 5.23a). In contrast, the ultralong hydroxyapatite nanowire paper was nonflammable and well preserved in fire (Figure 5.23b). Owing to the high thermal stability and excellent fire resistance of ultralong hydroxyapatite nanowires, the secret paper did not collapse during the decryption process with fire as a stimulus. Edible white vinegar purchased from the supermarket was used as an invisible ink to write on the secret paper. The pattern written with white vinegar on the secret paper was totally invisible after drying (Figure 5.23c and d); thus the information could be encrypted. Upon exposure to fire, the areas without writing on the white secret paper became shallow black due to the formation of the combustion residue from plant fibers. In contrast, the areas with writing using

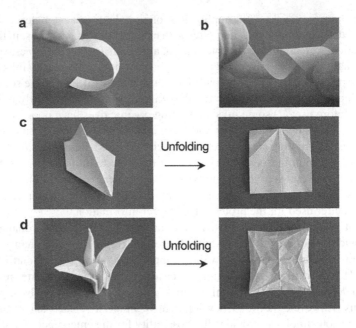

FIGURE 5.22   Digital images of the secret paper with excellent flexibility and processibility. (Reprinted with permission from reference [78])

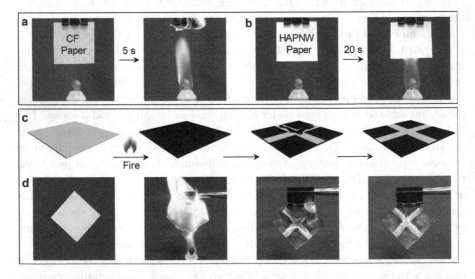

FIGURE 5.23   (a, b) Fire-resistance tests: (a) the plant fiber paper and (b) the ultralong hydroxyapatite nanowire fire-resistant paper; (c, d) schematic illustration (c) and digital images (d) of the color-changing process of the secret paper with covert information written using white vinegar in response to fire. (Reprinted with permission from reference [78])

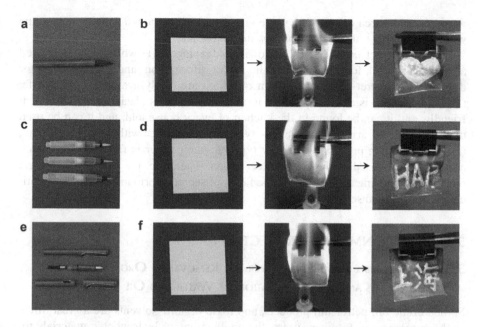

FIGURE 5.24 (a) Chinese writing brush; (b) color-changing (decryption) process of the secret paper using a Chinese writing brush as a writing tool and white vinegar as the invisible ink; (c) fountain pens with brush heads; (d) color-changing process of the secret paper using a fountain pen with a brush head as a writing tool and white vinegar as the invisible ink; (e) fountain pens with metal heads; (f) color-changing process of the secret paper using a fountain pen with a metal head as a writing tool and white vinegar as the invisible ink. (Reprinted with permission from reference [78])

white vinegar exhibited a gray-white color that could be distinguished clearly from the black areas without writing. As a result, the information written using white vinegar on the secret paper was revealed and readable, and the decryption process was completed. The color change was rapid and within 10 seconds in response to fire. By using various pens filled with white vinegar, you can easily write the secret information in a variety of ways on the new type of secret paper, as shown in Figure 5.24.

The principle of the encryption and decryption technology of the new kind of secret paper can be explained as follows: the new type of secret paper was made of ultralong hydroxyapatite nanowires and plant fibers. The ultralong hydroxyapatite nanowires were resistant to fire and high temperatures, while plant fibers could burn easily in fire. In the decryption process of the secret information on the secret paper in fire, the plant fibers were partially oxidized and burned to form black carbon particles, which covered the surface of the ultralong hydroxyapatite nanowires, making the color of the secret paper change from white to shallow black. Because of the catalytic effect of acetic acid molecules in white vinegar, the plant fibers in the area written with white vinegar could accelerate the combustion and oxidation to generate carbon dioxide gas and be removed, and essentially no or less black carbon particles were left on the area written with white vinegar which remained white (owing to the white color of the ultralong hydroxyapatite nanowires).

The new secret paper has the following advantages: (1) white vinegar is used as the security ink for encryption of the secret information, and fire is the key for decryption of secret information, both of which are easily available in daily life; (2) white vinegar is a common cooking material which is cheap and environment friendly, and it can be found in the kitchen of every household, and it can be used directly as the security ink of the new type of secret paper without any treatment; (3) the decryption process of the new type of secret paper is fast and convenient, and only fire is used as the key for decryption. Therefore, the new type of secret paper has promising applications in the fields of secret information protection, anti-counterfeiting, and security.[78]

5.5. ENVIRONMENTAL PROTECTION

5.5.1. RECYCLABLE ADSORPTION PAPER FOR REMOVAL OF ORGANIC SOLVENTS AND RAPID SEPARATION OF WATER AND OIL

Spilled crude oil, petroleum products, and toxic organic solvents are a great threat to the environment. In recent years, the applications of hydrophobic materials for treating these organic compounds were increasingly investigated. However, the practical applications of these materials have been hindered by their drawbacks such as low adsorption capacity, poor recyclability, and toxicity. Therefore, new high-performance adsorbents that can effectively and safely tackle these problems are highly desirable.

As discussed in Section 4.2, the as-prepared waterproof fire-resistant paper exhibited an excellent superhydrophobicity, and could be used as a highly effective adsorbent for a variety of organic compounds and oil-water separation, as shown in Figure 5.25.[59] The waterproof fire-resistant paper possessed a superoleophilic performance (Figure 5.25a), and it is promising as a highly efficient adsorbent for the removal of oil from water. As shown in Figure 5.25b, cyclohexane was dropped onto the water surface in a culture dish, and a piece of waterproof fire-resistant paper was put on the cyclohexane (Figure 5.25c). The cyclohexane could be completely adsorbed by the waterproof fire-resistant paper rapidly, and there was no residual cyclohexane on water after the oil–water separation (Figure 5.25d). The rapid and highly efficient oil–water separation performance could be attributed to the abundant pores, good capillary action, and excellent superhydrophobic property of the waterproof fire-resistant paper. Thus, the waterproof fire-resistant paper can be used for efficient treatment of water pollution resulting from oil spillage and industrial discharge of organic pollutants.

The fire-resistant inorganic paper consisting of ultralong hydroxyapatite nanowires was also investigated as a high-performance adsorbent for various organic pollutants. The experiments showed that the fire-resistant inorganic paper exhibited high adsorption capacities for various organic pollutants, as shown in Figure 5.26a, for instance, the fire-resistant inorganic paper had a high adsorption capacity of 7.3

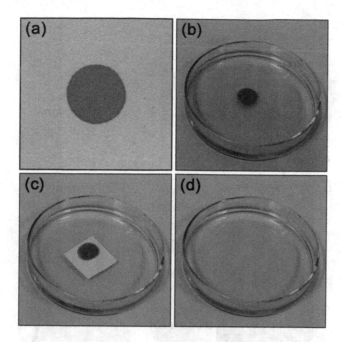

FIGURE 5.25 Oil–water separation tests of the waterproof fire-resistant paper based on ultralong hydroxyapatite nanowires. (a) Cyclohexane droplet (dyed with oil red) on the surface of the waterproof fire-resistant paper; (b) cyclohexane floating on the water surface; (c) cyclohexane was completely adsorbed by the waterproof fire-resistant paper; (d) cyclohexane was completely and rapidly removed by taking out the waterproof fire-resistant paper from the water surface, and the oil–water separation could be achieved. (Reprinted with permission from reference [59])

$g\ g^{-1}$ for chloroform. The recovery of pollutants and regeneration of used adsorbents are desirable for lowering the resource consumption and economic cost. The fire-resistant inorganic paper with adsorbed organic compounds could be easily recovered and regenerated by heating in a distillation unit or by burning in fire, and the adsorbed organic solvents could be completely removed from the fire-resistant inorganic paper for recycling. No visible damage to the fire-resistant inorganic paper was observed after many cycles of organic solvent removal. The regenerated fire-resistant inorganic paper could be reused for the adsorption of organic pollutants and exhibited similar adsorption capacities compared with the freshly prepared fire-resistant inorganic paper (Figure 5.26b). For most organic polymer adsorbents, the solvents and adsorbents cannot be recovered by heating owing to their poor thermal stability; only chemical extraction can be adopted, which makes the complete recovery of organic solvents difficult and is high cost. Furthermore, polymeric adsorbents may cause pollution to the environment. The fire-resistant inorganic paper consisting of ultralong hydroxyapatite nanowires has a high biocompatibility and is environment friendly, and possesses excellent recyclability, and has promising applications in treating various organic pollutants.[12]

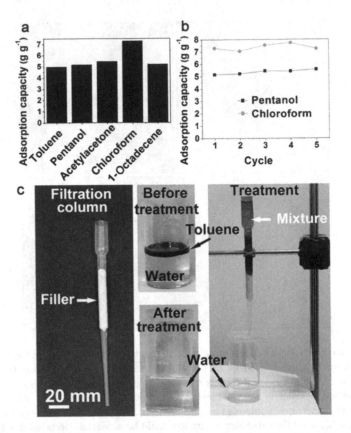

FIGURE 5.26 Adsorption properties for organic compounds of the fire-resistant inorganic paper consisting of ultralong hydroxyapatite nanowires. (a) Adsorption capacities for various organic compounds; (b) recycling performance for adsorbing pentanol and chloroform; (c) separation of a mixture of toluene (dyed by oil red O) and water by the filtration column using the fire-resistant inorganic paper as the filler. (Reprinted with permission from reference [12])

The fire-resistant inorganic paper consisting of ultralong hydroxyapatite nanowires could also be used as a filler to make a filtration column for the treatment of a mixture of organic pollutants and water.[12] The separation of a mixture of toluene (dyed by oil red O) and water is demonstrated in Figure 5.26c. When a mixture of toluene and water was added into the filtration column, water went through the fire-resistant inorganic paper filler, whereas toluene was adsorbed completely by the filler. By this simple and rapid procedure, toluene could be completely separated from water and the polluted water was effectively purified.

The superhydrophobic magnetic fire-resistant paper discussed in Section 4.5 can be used as a filter paper for oil–water separation.[62] As shown in Figure 5.27a, chloroform dyed with oil red could permeate rapidly through the superhydrophobic magnetic fire-resistant paper solely by gravity, while water could not go through the filter paper and stay on the upper surface of the filter paper. The separation efficiency was as high as

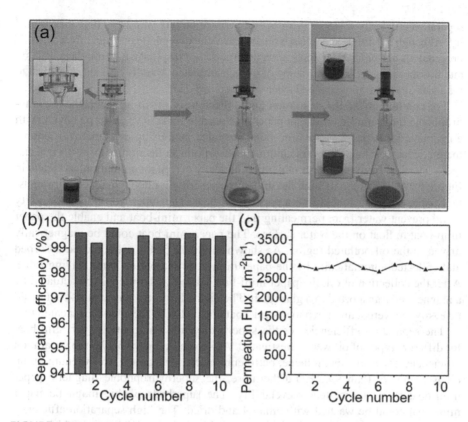

FIGURE 5.27 (a) Digital images of the oil (chloroform dyed with oil red)–water (dyed with methylene blue) separation process using the superhydrophobic magnetic fire-resistant paper as the filter paper; (b, c) recycling performance for oil (chloroform)–water separation of the superhydrophobic magnetic fire-resistant paper: (b) separation efficiencies for 10 cycles, and (c) permeation fluxes for 10 cycles. (Reprinted with permission from reference [62])

99.6%, implying the excellent oil–water separation performance of the superhydrophobic magnetic fire-resistant paper. The superhydrophobic magnetic fire-resistant paper could be washed using ethanol and recycled, and the separation efficiencies were higher than 99.0% after 10 cycles (Figure 5.27b). Benefiting from the highly porous structure formed by ultralong hydroxyapatite nanowires, the superhydrophobic magnetic fire-resistant paper showed a high permeation flux of 2835.9 L m^{-2} h^{-1}, and the high permeation flux could be well maintained after 10 cycles (Figure 5.27c), indicating the excellent recyclability of the superhydrophobic magnetic fire-resistant paper.

In another experiment, soil was dispersed in an oil–water mixture. Then, the superhydrophobic magnetic fire-resistant paper was used as a filter paper to separate the oil from the water containing soil. The experimental results indicated that oil quickly went through the superhydrophobic magnetic fire-resistant paper, and the soil and water were retained above the superhydrophobic magnetic fire-resistant paper. Thus, clean oil was successfully separated and collected from the oil–water mixture containing soil. The separation efficiency was 97.8 ± 1.1%, and permeation flux was 2,034.6 ± 262.2 L m^{-2} h^{-1}. Although the filtration process is an effective way to separate oil from oily water, this procedure is not suitable for application in large-scale oil spillage because the collection and transportation of a large volume of oil-contaminated water is difficult. It is ideal to realize *in situ* separation and collection of the spilled oil.[62]

For continuous *in situ* separation and collection of oil from water, the superhydrophobic magnetic fire-resistant paper was used to make a vessel-type device with a magnetic function. A paper mini-boat was fabricated using the superhydrophobic magnetic fire-resistant paper for automatic absorption, collection, transportation, and recovery of oil from oily water under magnetically driven manipulation, as shown in Figure 5.28a. Owing to its high flexibility, the superhydrophobic magnetic fire-resistant paper could be folded into a paper mini-boat. The superhydrophobic property could prevent water from permeating into the paper mini-boat and enable the paper mini-boat to float on the water surface. The paper mini-boat could be magnetically driven to the oil-polluted region. As shown in Figure 5.28b, oil could be absorbed and collected automatically by the paper mini-boat without an extra driving force. After the collection of oil, the paper mini-boat loaded with oil could be actuated by a magnet and transported to a glass dish. The oil loaded in the paper mini-boat could be easily recovered, and the paper mini-boat could also be recycled for repeated use.

The separation efficiencies of the superhydrophobic magnetic paper mini-boat for different types of oil were investigated. As shown in Figure 5.29a, the separation efficiencies for isooctane, toluene, petroleum ether, soybean oil, and vacuum pump oil were higher than 99.2%. Furthermore, the superhydrophobic magnetic paper mini-boat showed excellent recyclability. The superhydrophobic magnetic paper mini-boat could be washed with ethanol and dried. The high separation efficiency of the superhydrophobic magnetic paper mini-boat for cyclohexane could be well maintained after 10 cycles (Figure 5.29b).[62]

5.5.2. WATER PURIFICATION AND WASTEWATER TREATMENT

Water is a source of life, and human society cannot survive and develop without water resources. In recent years, because of increasingly severe water pollution worldwide, water shortages and water crises have become a global problem. The amount of fresh water on Earth is less than 3% of the total water and is much lesser than that of seawater. In addition, more than 70% of the freshwater is frozen in severe cold zones, which is difficult for humans to use.

The most promising way to solve the severe freshwater crisis is to utilize water that is not directly drinkable, such as wastewater and seawater. To develop various water treatment technologies or new filter materials is of great significance to alleviating the water crisis. Among them, the membrane separation technology has the advantages of simple operation, low equipment requirements, and high separation efficiency, and has become one of the most important methods in separation science today. Filtration membranes are essentially microporous barriers of polymeric, ceramic, metallic, or other materials which are used to separate dissolved materials (solutes), colloids, or fine particulates from solutions. Membranes are generally classified into four categories based on their

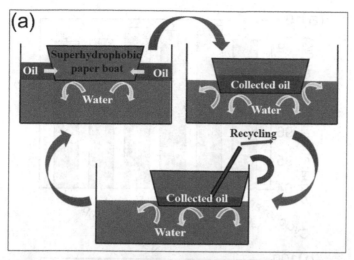

FIGURE 5.28 Automatic absorption, collection, transportation, and recovery of oil from oily water using the superhydrophobic magnetic fire-resistant paper under the magnetically driven manipulation. (a) Schematic illustration of the superhydrophobic magnetic paper mini-boat made from the superhydrophobic magnetic fire-resistant paper for absorption, collection, transportation, and recovery of oil from oily water under the magnetically driven manipulation; (b) digital images of the process for selective oil absorption, collection, transportation, and recovery of oil from oily water by the superhydrophobic magnetic fire-resistant paper mini-boat using a magnet. (Reprinted with permission from reference [62])

mean pore size: (1) hyperfiltration or reverse osmosis usually separates molecules or ions with sizes less than 1 nm, such as monovalent ions from water; (2) nanofiltration separates molecules or ions with sizes of 1 nm up to several nanometers, such as sugars and divalent ions, while allowing the passage of monovalent ions; (3) ultrafiltration can separate materials in the range of 1–100 nm, such as proteins or colloids; (4)

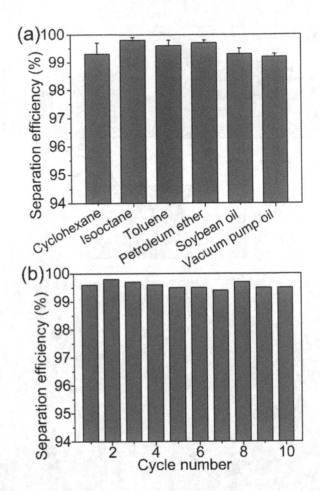

FIGURE 5.29 (a) Separation efficiencies for various organic compounds of the superhy-drophobic magnetic paper mini-boat made from the superhydrophobic magnetic fire-resis-tant paper; (b) separation efficiencies of the superhydrophobic magnetic paper mini-boat for cyclohexene for 10 cycles. (Reprinted with permission from reference [62])

microfiltration can be used for sterilization by removing insoluble particulate materials (microbes) ranging from 0.1–10 μm.[79]

Although various commercial water filtration membranes are widely used in indus-try and daily life, there are still some problems that need to be solved. For example, some filter membrane materials have disadvantages such as poor biocompatibility, use of harmful chemicals in the preparation process, and unsatisfactory separation efficiency. In addition, many filter membranes have a single function. In practical applications, it is often necessary to use the multi-stage filter system to improve the efficiency of water treatment, but this will undoubtedly make the water treatment equipment more compli-cated and increase the cost. Therefore, environment friendly, multifunctional, and highly efficient filter materials have attracted great attention.

(1) Microfiltration Filter Paper Based on Ultralong Hydroxyapatite Nanowires

The author's research team developed a new kind of microfiltration filter paper using ultralong hydroxyapatite nanowires as the main construction material, natural plant fibers as the additive, and polyamidoamine epichlorohydrin resin as a reinforcement to increase the wet mechanical strength of the filter paper.[80] The combination of ultralong hydroxyapatite nanowires and natural plant fibers could obviously improve the mechanical strength of the microfiltration filter paper, and the addition of polyamidoamine epichlorohydrin resin could enhance the wet mechanical strength of the microfiltration filter paper. The ultralong hydroxyapatite nanowires were interweaved with each other to form a nanoscale porous network structure, which could significantly increase the porosity of the filter paper, providing the filter paper with superhydrophilicity, and interception and adsorption capabilities for pollutants. The as-prepared microfiltration filter paper exhibited excellent separation performance and could effectively remove various pollutants in water, and its removal efficiency was high. The good adsorption and ion exchange properties of ultralong hydroxyapatite nanowires enabled the filter paper to effectively remove organic dyes and heavy metal ions. The new type of microfiltration filter paper exhibited both excellent filtration and adsorption properties, and could be used repeatedly and for a long time. The microfiltration filter paper holds promise for applications in high-performance water purification, clean water regeneration, and other fields.

Figure 5.30a-f shows digital images of the as-prepared new kind of microfiltration filter paper sheets with different contents of ultralong hydroxyapatite nanowires and a diameter of 20 cm, and the paper sheets showed a uniform white color. Many filtration membranes reported in the literature were small with sizes of a few centimeters. The preparation of the large-sized filter paper can broaden its application field and reduce the preparation cost. The author's research group successfully prepared a large-sized microfiltration filter paper sheet with a length of 44 cm and a width of 31 cm, which is promising for application in water treatment (Figure 5.30g).

Plant fibers are the primary raw material in traditional papermaking, and the wet strength agent is a common additive in paper. Polyamidoamine epichlorohydrin resin is a water-soluble, cationic thermosetting resin which is an excellent wet strength agent in the paper industry. Compared with many traditional wet strength agents, polyamidoamine epichlorohydrin has advantages such as nontoxicity, wide application range, small addition amount, and superior reinforcing effect. Furthermore, polyamidoamine epichlorohydrin is a non-formaldehyde wet strength agent, and can be applied in the filter paper, cosmetic paper, food wrapping paper, and medical paper. The addition of the polyamidoamine epichlorohydrin resin enabled the microfiltration filter paper to maintain a relatively high mechanical strength in the wet state. The experiments showed that the as-prepared microfiltration filter paper exhibited superior mechanical properties. The tensile strength of the dried microfiltration filter paper was higher than that of the wet microfiltration filter paper. The highest wet tensile strength of the as-prepared microfiltration filter paper was observed when the addition amount of polyamidoamine epichlorohydrin was 3 wt.%. The tensile strengths of the microfiltration filter paper sheets with different contents of ultralong hydroxyapatite nanowires were measured, and it was found that the tensile strength

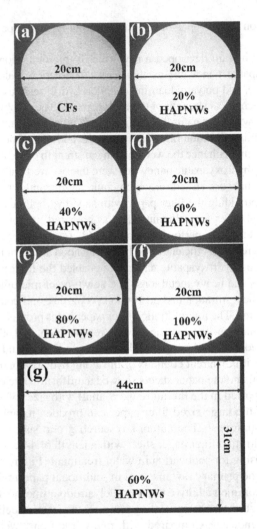

FIGURE 5.30 Digital images of the as-prepared new kind of microfiltration filter paper sheets consisting of ultralong hydroxyapatite nanowires and plant fibers with different weight ratios of hydroxyapatite/(hydroxyapatite + plant fibers): (a) 0 wt.% (100 wt.% plant fibers); (b) 20 wt.%; (c) 40 wt.%; (d) 60 wt.%; (e) 80 wt.%; (f) 100 wt.%; (g) a large-sized microfiltration filter paper sheet with 60 wt.% ultralong hydroxyapatite nanowires. (Reprinted with permission from reference [80])

decreased with increasing amount of ultralong hydroxyapatite nanowires for both dried and wet microfiltration filter paper sheets. The tensile strengths of the dried microfiltration filter paper sheets were 41.32 MPa, 18.14 MPa, 10.02 MPa, 6.05 MPa, and 1.58 MPa, respectively, for the weight ratio of ultralong hydroxyapatite nanowires of 20 wt.%, 40 wt.%, 60 wt.%, 80 wt.%, and 100 wt.%, and the tensile strengths of the wet microfiltration filter paper sheets were 18.70 MPa, 10.42 MPa, 8.05 MPa, 2.09 MPa, and 0.71 MPa, respectively. The wet tensile strength could retain 80.3%

of that of the dried microfiltration filter paper when the weight ratio of ultralong hydroxyapatite nanowires was 60 wt.%.[80]

The physical properties of the microfiltration filter paper sheets with different weight ratios of ultralong hydroxyapatite nanowires (0 wt.%, 20 wt.%, 40 wt.%, 60 wt.%, 80 wt.%, 100 wt.%) were investigated. With increasing content of ultralong hydroxy-apatite nanowires, the basis weight of the microfiltration filter paper decreased from 79.33 g m^{-2} to 64.76 g m^{-2}, the thickness increased from 101 μm to 158 μm, the bulk increased from 1.27 cm^3 g^{-1} to 2.44 cm^3 g^{-1}, and the porosity increased from 28.96% to 66.20%. Although the polyamidoamine epichlorohydrin resin could enhance the hydrophobicity of the microfiltration filter paper, the hydrophilicity and water uptake of the microfiltration filter paper could be greatly increased as the weight ratio of ultralong hydroxyapatite nanowires increased. The water contact angle decreased from 88.40° to 0° and the water absorption capacity increased when the weight ratio of ultralong hydroxyapatite nanowires increased from 0% to 100%.

The pore size distribution of the as-prepared microfiltration filter paper was evaluated by the bubble point method. The average pore diameters of the microfiltration filter paper sheets were 117.5 nm, 135.2 nm, 139.0 nm, and 144.3 nm for the weight ratio of ultralong hydroxyapatite nanowires of 20 wt.%, 40 wt.%, 60 wt.% and 80 wt.%, respectively. The average pore size of the cellulose fiber filter paper with 3 wt.% polyamidoamine epichlo-rohydrin was as small as 17.2 nm. By the addition of ultralong hydroxyapatite nanowires, the pore size of the microfiltration filter paper could be increased. Another advantage of the microfiltration filter paper was that the pore size distribution was relatively narrow. The interactions between the fibers played an important role in determining the pore size distribution. Cellulose fibers were interconnected by a large number of hydrogen bonds. Ultralong hydroxyapatite nanowires and cellulose fibers interweaved with each other by the van der Waals force and hydrogen bonds at the interfaces. As the weight ratio of ultralong hydroxyapatite nanowires increased, the interactions between cellulose fibers decreased, resulting in an increase in the average pore size.[80]

The filtration performance of the as-prepared microfiltration filter paper was eval-uated by a cross-flow low-pressure flat membrane test equipment with an effective diameter of 7.8 cm, as shown in Figure 5.31a. The pure water flux as a function of working pressure of the as-prepared microfiltration filter paper sheets with different weight ratios of ultralong hydroxyapatite nanowires is shown in Figure 5.31b. Usually, the pure water flux is determined by the porosity of the filter paper. As for the plant fiber filter paper, the pure water flux was only 4.87 L m^{-2} h^{-1} under a pressure of 1 bar. When the polyamidoamine epichlorohydrin resin was added into the filter paper, the pure water flux of the plant fiber filter paper decreased to 0.09 L m^{-2} h^{-1} due to its low porosity (28.96%). When the weight ratio of ultralong hydroxyapatite nanowires increased from 20 wt.% to 80 wt.%, the pure water flux of the microfiltration filter paper enhanced from 18.12 L m^{-2} h^{-1} to 287.28 L m^{-2} h^{-1} at a pressure of 1 bar. In addi-tion, the pure water flux of the microfiltration filter paper increased with increasing working pressure (Figure 5.31b). The enhanced pure water flux of the microfiltration filter paper was attributed to its highly porous structure and superhydrophilicity.

An aqueous suspension containing TiO$_2$ nanoparticles with an average particle size of ~40 nm was adopted to evaluate the filtration performance of the as-prepared

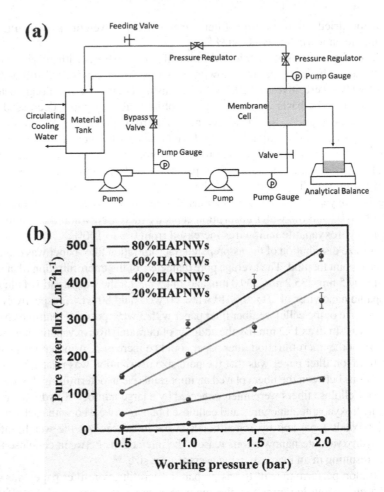

FIGURE 5.31 (a) Schematic diagram of the cross-flow low-pressure flat membrane test equipment; (b) pure water flux versus working pressure of the as-prepared microfiltration filter paper sheets with different weight ratios of ultralong hydroxyapatite nanowires. (Reprinted with permission from reference [80])

microfiltration filter paper. Figure 5.32a shows that the rejection percentages of the microfiltration filter paper sheets with different weight ratios of ultralong hydroxyapatite nanowires were higher than 99.86% for TiO_2 nanoparticles at a concentration of 250 ppm. Furthermore, the rejection percentages of the microfiltration filter paper with 60 wt.% ultralong hydroxyapatite nanowires for TiO_2 nanoparticles at different concentrations from 50 ppm to 500 ppm were higher than 98.61% (Figure 5.32b).

The water filtration experiments were performed continuously for 4 hours to evaluate the water filtration stability of the microfiltration filter paper. Figure 5.32c shows the water fluxes and rejection percentages of TiO_2 nanoparticles at a concentration of 250 ppm during a time period of 4 hours. The water flux and rejection percentage for TiO_2 nanoparticles reduced to some extent during 4-hour filtration process because

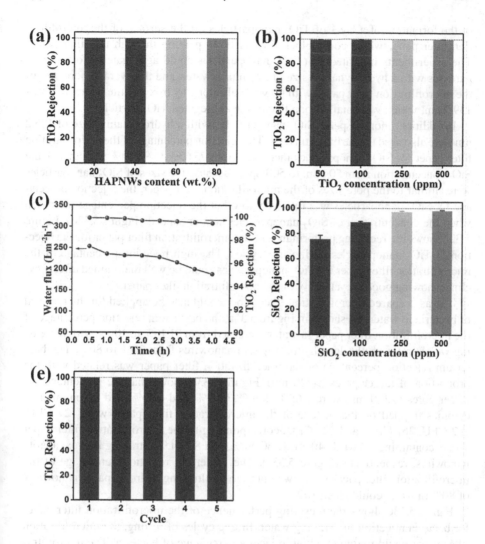

FIGURE 5.32 Particle rejection properties of the as-prepared microfiltration filter paper sheets with different weight ratios of ultralong hydroxyapatite nanowires. (a) Rejection percentage of TiO_2 nanoparticles at a concentration of 250 ppm versus weight ratio of ultralong hydroxyapatite nanowires; (b) rejection percentages of TiO_2 nanoparticles at different concentrations using the microfiltration filter paper with 60 wt.% ultralong hydroxyapatite nanowires; (c) water fluxes and rejection percentages of TiO_2 nanoparticles at a concentration of 250 ppm during a time period of four hours using the microfiltration filter paper with 60 wt.% ultralong hydroxyapatite nanowires; (d) rejection percentages of SiO_2 nanoparticles at different concentrations using the microfiltration filter paper with 60 wt.% ultralong hydroxyapatite nanowires; (e) recycling performance of the microfiltration filter paper for rejection of TiO_2 nanoparticles. All of the experiments were carried out at a working pressure of 1 bar. (Reprinted with permission from reference [80])

of the formation of a layer of TiO_2 nanoparticles on the surface of the microfiltration filter paper which could block tiny particles passing through the filter paper. The experiments indicated that TiO_2 nanoparticles could aggregate to form larger particles with a hydrodynamic size of 321 nm in water, and the average pore size of the microfiltration filter paper with 60 wt.% ultralong hydroxyapatite nanowires was 139.0 nm, which was smaller than that of the aggregated TiO_2 particles.

In addition, monodisperse SiO_2 nanoparticles with a hydrodynamic size of 185.2 nm were also used for the filtration tests. The rejection percentage of the microfiltration filter paper for SiO_2 nanoparticles increased from 75.65% to 98.04% with increasing SiO_2 concentration from 50 ppm to 500 ppm. Because the sizes of SiO_2 nanoparticles were similar to the pore sizes of the microfiltration filter paper, the rejection percentage for SiO_2 nanoparticles was relatively low, but the rejection percentage increased when the concentration of SiO_2 nanoparticles was enhanced (Figure 5.32d). Figure 5.32e shows the recycling performance of the microfiltration filter paper for the rejection of TiO_2 nanoparticles during five cycles. The high rejection percentage of the microfiltration filter paper for TiO_2 nanoparticles could be well maintained after recycling, showing good recyclability of the microfiltration filter paper.[80]

The as-prepared microfiltration filter paper could also be applied for the removal of bacteria in water. As shown in Figure 5.33a, the bacterium rejection percentage of the microfiltration filter paper in water increased from 91.16% to 100% with increasing weight ratio of ultralong hydroxyapatite nanowires from 20% to 80%. The bacterium rejection percentage of the microfiltration filter paper was related with the proportion of larger pores (>220 nm). Figure 5.33b shows that the percentages of larger pores (>220 nm) were 1.68%, 1.30%, 0.49%, and 0.53%, and the number of colonies trapped on the surface of the microfiltration filter paper were 127 CFU, 182 CFU, 361 CFU, and 229 CFU, corresponding to the microfiltration filter paper sheets containing 20 wt.%, 40 wt.%, 60 wt.%, and 80 wt.% ultralong hydroxyapatite nanowires, respectively (Figure 5.33b). The bacterium rejection percentage of the microfiltration filter paper with a weight ratio of ultralong hydroxyapatite nanowires of 80% in water could reach 100%.

Figure 5.33c shows the recycling performance of the microfiltration filter paper for bacteria rejection in drinking water. In five cycles of testing, the microfiltration filter paper could maintain a high rejection percentage of bacteria. The microfiltration filter paper possessed a porous network structure constructed by interwoven ultralong hydroxyapatite nanowires and plant cellulose fibers, and ultralong hydroxyapatite nanowires filled in the pores formed by cellulose fibers. The size exclusion and blocking effect of the microfiltration filter paper could effectively trap particles and bacteria in water (Figure 5.33d).

The adsorption properties of the microfiltration filter paper with 60 wt.% ultralong hydroxyapatite nanowires for the organic dye and heavy metal ions were also investigated. Figure 5.34a shows the adsorption performance of the microfiltration filter paper for methyl blue. The adsorption capacity (Q_e) for methyl blue of the microfiltration filter paper increased with increasing equilibrium concentration (C_e), and the obvious color change from white to blue of the microfiltration filter paper (the inset) showed high-performance adsorption of methyl blue. The Langmuir and

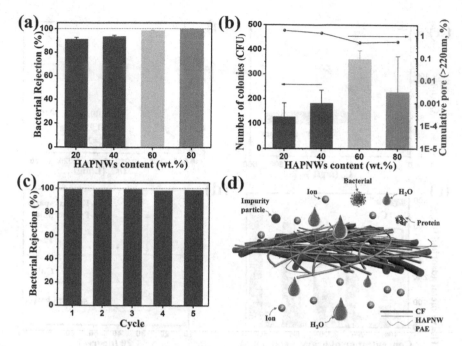

FIGURE 5.33 Bacterium rejection performance of the as-prepared microfiltration filter paper sheets with different weight ratios of ultralong hydroxyapatite nanowires. (a) Rejection percentages versus weight ratio of ultralong hydroxyapatite nanowires for bacteria in drinking water; (b) number of bacterium colonies trapped on the surface of the microfiltration filter paper after filtration versus weight ratio of ultralong hydroxyapatite nanowires, and the cumulative percentages of pores larger than 220 nm in the microfiltration filter paper measured by the bubble point method; (c) recycling performance of the microfiltration filter paper for the rejection of bacteria in drinking water; (d) schematic illustration of the structure and filtration process of the microfiltration filter paper. All of the experiments were carried out at a working pressure of 1 bar. (Reprinted with permission from reference [80])

Freundlich models were adopted to analyze the adsorption of methyl blue by the microfiltration filter paper. It was found that the linear regression coefficient (R^2) of the Langmuir equation was higher than that of the Freundlich equation, indicating that the Langmuir adsorption model could provide a better explanation for methyl blue adsorption of the microfiltration filter paper. The maximum adsorption capacity (Q_{max}) obtained from the Langmuir equation was 273.97 mg g^{-1} (Figure 5.34b).

The microfiltration filter paper based on ultralong hydroxyapatite nanowires could also be used for the removal of heavy metal ions in water. Figure 5.34c shows the adsorption amounts of the microfiltration filter paper with 60 wt.% ultralong hydroxyapatite nanowires for Cu^{2+} and Pb^{2+} ions with different initial concentrations. The adsorption amount of the microfiltration filter paper for Cu^{2+} ions at an initial concentration of 500 ppm was 60.42 mg g^{-1}. The adsorption amount of the microfiltration filter paper for Pb^{2+} ions almost linearly increased with the initial concentration of Pb^{2+} ions, and reached as high as 508.16 mg g^{-1} at an initial concentration of

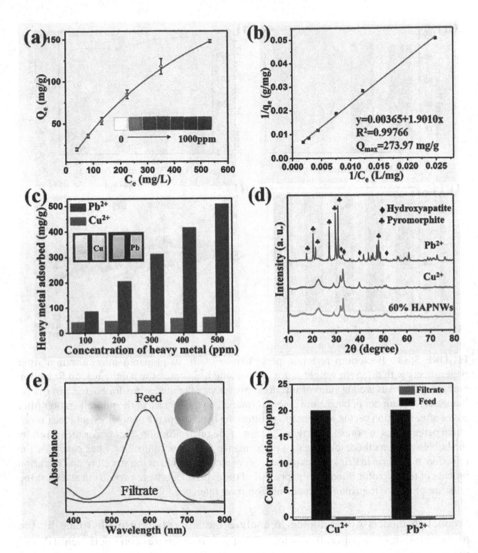

FIGURE 5.34 Adsorption properties for the organic dye and heavy metal ions of the microfiltration filter paper with 60 wt.% ultralong hydroxyapatite nanowires. (a) Adsorption isotherms for methyl blue; (b) Langmuir model for methyl blue adsorption; (c) adsorption amount of Cu^{2+} ions or Pb^{2+} ions in the $CuCl_2$ or $PbCl_2$ aqueous solutions with different initial concentrations; (d) XRD patterns of the microfiltration filter paper before and after adsorption tests for heavy metal ions at an initial concentration of 500 ppm for 48 hours; (e) UV–visible absorption spectra of the feeding and filtrate solutions, and digital images of the microfiltration filter paper (inset) before and after filtration of methyl blue aqueous solution with an initial concentration of 20 ppm; (f) adsorption performance of the microfiltration filter paper for Cu^{2+} ions or Pb^{2+} ions with a concentration of 20 ppm under the filtration mode. (Reprinted with permission from reference [80])

500 ppm with an adsorption efficiency of 98.9%. The XRD patterns of the micro-filtration filter paper before and after adsorption are shown in Figure 5.34d. It was found that pyromorphite ($Pb_5(PO_4)_3Cl$) was formed after the adsorption of Pb^{2+} ions in the microfiltration filter paper, and that the adsorption of Pb^{2+} ions by the micro-filtration filter paper was accompanied by the obvious release of Ca^{2+} ions in the aqueous solution. The weight of the microfiltration filter paper increased after the adsorption of Pb^{2+} ions and with increasing initial Pb^{2+} ion concentration. The XRD pattern of the microfiltration filter paper had no obvious change after adsorption of Cu^{2+} ions, and the weight of the microfiltration filter paper had no obvious change after the adsorption of Cu^{2+} ions. The experimental results showed that the adsorp-tion process was dominated by the ion adsorption mechanism in the case of Cu^{2+} ions. In contrast, the ion exchange dominated and a stable crystal phase of pyromor-phite formed in the case of Pb^{2+} ions.[80]

In the filtration mode, the microfiltration filter paper exhibited good adsorption performance for dyes and heavy metal ions. After an aqueous solution (100 mL) of methyl blue with an initial concentration of 20 ppm was filtered through the micro-filtration filter paper with 60 wt.% ultralong hydroxyapatite nanowires, the surface of the microfiltration filter paper showed a dark blue color, and the adsorption effi-ciency for methyl blue was 96.39% (Figure 5.34e). In addition, the microfiltration filter paper exhibited a good adsorption capacity for a low-concentration aqueous solution of Cu^{2+} or Pb^{2+} ions under dynamic conditions, and the removal efficiency was 100% at an initial ion concentration of 20 ppm (Figure 5.34f).

Usually, excellent hydrophilicity, electrical neutrality, and no hydrogen-bonding donors are prerequisites for superior antifouling performance of the filter paper. Although the added polyamidoamine epichlorohydrin could decrease the hydrophi-licity of the microfiltration filter paper, ultralong hydroxyapatite nanowires could reduce the contact angle to below 5°. Figure 5.35a shows the water flux of the micro-filtration filter paper with 60 wt.% ultralong hydroxyapatite nanowires before and after the filtration of an aqueous solution containing bovine serum albumin. When the feeding solution changed from pure water to bovine serum albumin aqueous solution, the water flux decreased from 206.6 L m^{-2} h^{-1} to 119.8 L m^{-2} h^{-1} within 1 hour, and the rejection percentage of bovine serum albumin was 17.21%. After the fouled microfiltration filter paper was immersed in pure water for 1 hour, the pure water flux increased from 119.8 L m^{-2} h^{-1} to 159.8 L m^{-2} h^{-1}. The flux recovery ratio (FRR), total flux decline ratio (Rt), reversible fouling ratio (Rr), and irreversible foul-ing ratio (Rir) of the microfiltration filter paper with 60 wt.% ultralong hydroxyapa-tite nanowires were 67.19%, 42.10%, 9.29%, and 32.81%, respectively (Figure 5.35b). The protein molecules were easily trapped and constrained in the pores of the micro-filtration filter paper. The environmentally friendly microfiltration filter paper based on ultralong hydroxyapatite nanowires with both excellent filtration and adsorption properties has promising applications in high-performance water purification to tackle the water scarcity problem worldwide.[80]

(2) Microfiltration Filter Paper with Regulated Pore Size by Double Metal Oxide Nanosheets for High-Performance Dye Separation

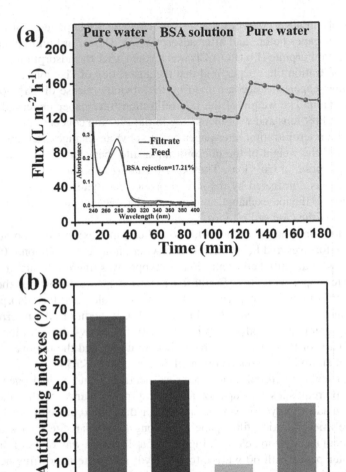

FIGURE 5.35 Antifouling performance of the microfiltration filter paper with 60 wt.% ultralong hydroxyapatite nanowires. (a) Water flux of the microfiltration filter paper and UV–visible absorption spectra of bovine serum albumin solution (inset) before and after filtration of the bovine serum albumin aqueous solution (500 ppm); (b) antifouling indexes of the microfiltration filter paper. (Reprinted with permission from reference [80])

Enhancements in both water flux and removal efficiency of separation membranes during the filtration process are still a big challenge. The porous structure of the filter paper can provide many pores and channels for water transportation, but the separation performance is usually poor. The author's research group developed a new kind of water purification microfiltration filter paper consisting of ultralong hydroxyapatite nanowires, cellulose fibers, and double metal oxide nanosheets to realize simultaneous enhancement of both water flux and removal efficiency for high-performance dye separation.[81] Positively charged double metal oxide nanosheets could adsorb on the surface of negatively charged ultralong hydroxyapatite nanowires and embed in the porous networked structure of the filter paper, which could provide a porous structure for rapid water transportation and adjust the pore size of the filter paper. As a result, the pure water flux of the microfiltration filter paper could be regulated. The optimized pure water flux of the microfiltration filter paper could reach 783.6 L m^{-2} h^{-1} bar^{-1}, which was 1.51 times that of the microfiltration filter paper without double metal oxide nanosheets (518.6 L m^{-2} h^{-1} bar^{-1}). The optimized rejection percentage and water flux of the microfiltration filter paper with double metal oxide nanosheets for Congo red were significantly enhanced (98.3% and 736.8 L m^{-2} h^{-1} bar^{-1}, respectively) compared with the microfiltration filter paper without double metal oxide nanosheets. The experiments indicated that the size of metal oxide nanosheets had a great effect on the water flux and dye rejection percentage of the microfiltration filter paper.

Double metal oxide nanosheets could embed in the porous networked structure formed by interwoven ultralong hydroxyapatite nanowires and cellulose fibers, and adjust the pore size of the microfiltration filter paper. When the weight ratio of double metal oxide nanosheets increased, the average pore size first increased and then decreased. The average pore size of the microfiltration filter paper without double metal oxide nanosheets was 115.3 nm, and the average pore size of the microfiltration filter paper sheets with double metal oxide nanosheets of 0.200, 0.400, and 0.600 g was 155.3, 158.5, and 117.3 nm, respectively. As a result, the pure water flux of the microfiltration filter paper could be adjusted. The pure water flux of the microfiltration filter paper with double metal oxide nanosheets of 0.400 g was as high as 783.6 L m^{-2} h^{-1} bar^{-1}, which was 1.51 times that of the microfiltration filter paper without double metal oxide nanosheets (518.6 L m^{-2} h^{-1} bar^{-1}). Thus, double metal oxide nanosheets could be applied as an efficient material to adjust the water flux of the filter paper.

The filtration performance of the microfiltration filter paper with double metal oxide nanosheets was investigated using different dyes. The experimental results showed that the addition of double metal oxide nanosheets could increase the removal efficiency of dyes. As shown in Figure 5.36a, the rejection percentage of Congo red by the microfiltration filter paper without double metal oxide nanosheets was only 59.8%, and its water flux was 534.7 L m^{-2} h^{-1} bar^{-1}. However, the rejection percentage and water flux of the microfiltration filter paper with 0.200 g double metal oxide nanosheets for Congo red greatly increased (94.4% and 718.4 L m^{-2} h^{-1} bar^{-1}, respectively). The rejection percentage for Congo red was 98.3%, and the water flux was 736.8 L m^{-2} h^{-1} bar^{-1} for the microfiltration filter paper with 0.400 g double metal oxide nanosheets. By further increasing the addition amount of double metal oxide nanosheets, the water flux of the microfiltration filter paper obviously reduced to 398.1 L m^{-2} h^{-1} bar^{-1} because water transporting channels were blocked by more double metal oxide nanosheets, although the rejection percentage for Congo red was further increased to 99.6%. The microfiltration filter paper with 0.400 g double metal oxide nanosheets exhibited the best filtration performance in terms of both rejection percentage and water flux. In addition, the red color on the back side of the filter paper gradually became light from the microfiltration filter paper

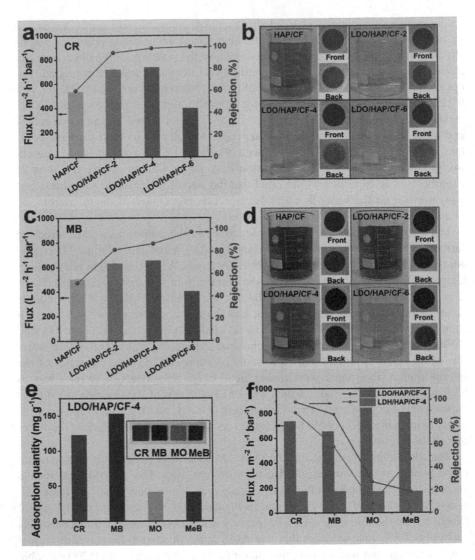

FIGURE 5.36 Filtration properties of the microfiltration filter paper with double metal oxide nanosheets for various dyes in comparison with the microfiltration filter paper with double metal hydroxide nanosheets. (a, b) Water fluxes and rejection percentages of the microfiltration filter paper sheets with different sizes of double metal oxide nanosheets obtained by the hydrothermal treatment for different times (12 hours, 24 hours, and 72 hours, respectively) for Congo red (CR), digital images of the filtrate solutions and filter paper sheets after filtration are shown in (b); (c, d) methyl blue (MB), digital images of the filtrate solutions and filter paper sheets after filtration are shown in (d); (e) adsorption amounts of the microfiltration filter paper with 0.400 g double metal oxide nanosheets for Congo red, methyl blue, methyl orange (MO), and methylene blue (MeB); (f) comparison of water flux and rejection percentage of the microfiltration filter paper with 0.400 g double metal oxide nanosheets and the microfiltration filter paper with 0.400 g double metal hydroxide nanosheets for Congo red, methyl blue, methyl orange, and methylene blue. All dye solutions had a concentration of 50 ppm and a volume of 100 mL. (Reprinted with permission from reference [81])

without double metal oxide nanosheets to the microfiltration filter paper with 0.600 g double metal oxide nanosheets, indicating that Congo red adsorption of the microfiltration filter paper was still not saturated (Figure 5.36b). The adsorption and filtration behaviors of the microfiltration filter paper with double metal oxide nanosheets for methyl blue, methyl orange, and methylene blue were similar to those of Congo red. As shown in Figure 5.36c, the microfiltration filter paper with 0.400 g double metal oxide nanosheets exhibited a water flux of 654.4 L m^{-2} h^{-1} bar^{-1} and a rejection percentage of 87.1% for methyl blue. The microfiltration filter paper with 0.400 g double metal oxide nanosheets had a water flux of 843.0 L m^{-2} h^{-1} bar^{-1} and 805.1 L m^{-2} h^{-1} bar^{-1}, respectively, for methyl orange and methylene blue, and a rejection percentage of 27.3% and 18.6%, respectively, for methyl orange and methylene blue. Furthermore, the microfiltration filter paper with double metal oxide nanosheets could be well preserved after filtration and adsorption of the dye (Figure 5.36d). The rejection percentage for the four dyes of the microfiltration filter paper with 0.400 g double metal oxide nanosheets was in the following order: Congo red > methyl blue > methyl orange > methylene blue.[81]

Figure 5.36e shows the adsorption amounts of the microfiltration filter paper with 0.400 g double metal oxide nanosheets for Congo red, methyl blue, methyl orange, and methylene blue, which were 122.7 mg g^{-1}, 153.1 mg g^{-1}, 41.5 mg g^{-1}, and 41.3 mg g^{-1}, respectively. Congo red, methyl blue, methyl orange dye molecules were negatively charged, and methylene blue molecules were positively charged. Because methylene blue molecules were positively charged, the adsorption and filtration processes of the microfiltration filter paper with double metal oxide nanosheets were less efficient. The adsorption of dye molecules by the microfiltration filter paper with double metal oxide nanosheets was mainly attributed to the electrostatic interaction between dye molecules and the filter paper, surface complexation between ultralong hydroxyapatite nanowires and dye molecules, and hydrogen bonding between cellulose fibers/ultralong hydroxyapatite nanowires and dye molecules, among others. The microfiltration filter paper with double metal oxide nanosheets exhibited the highest adsorption amount for methyl blue, probably due to the hydrogen bonding between the sulfonic acid groups in dye molecules and cellulose fibers/ultralong hydroxyapatite nanowires, and higher molecular weight of methyl blue.

In addition, the charge of nanosheets had an effect on the filtration of the dye. The double metal hydroxide nanosheets were also adopted for comparison. As shown in Figure 5.36f, the average water flux of the microfiltration filter paper with 0.400 g double metal hydroxide nanosheets for dye aqueous solutions was 171.2 L m^{-2} h^{-1} bar^{-1}, which was much lower than that of the microfiltration filter paper with double metal oxide nanosheets. Furthermore, the rejection percentages of the microfiltration filter paper with 0.400 g double metal hydroxide nanosheets for Congo red, methyl blue, and methyl orange were significantly lower than those of the microfiltration filter paper with 0.400 g double metal oxide nanosheets.

The size effect of double metal oxide nanosheets on the filtration performance of the microfiltration filter paper was investigated. As shown in Figure 5.37a–c, the average hydrodynamic size of double metal oxide nanosheets prepared for a hydrothermal reaction time of 12 hours, 24 hours, and 72 hours was 369.7 nm, 464.4 nm,

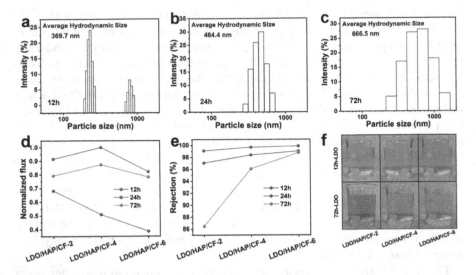

FIGURE 5.37 Filtration properties of the microfiltration filter paper sheets with different sizes of double metal oxide nanosheets prepared by the hydrothermal treatment for different times (12 hours, 24 hours, and 72 hours, respectively). (a–c) Size distributions of double metal oxide nanosheets prepared by the hydrothermal treatment for different times; (d) normalized water fluxes and (e) rejection percentages for Congo red of the microfiltration filter paper sheets with different sizes of double metal oxide nanosheets; (f) digital images of the Congo red filtrate using the microfiltration filter paper sheets with different sizes of double metal oxide nanosheets. All dye solutions had a concentration of 50 ppm and a volume of 100 mL. (Reprinted with permission from reference [81])

and 666.5 nm, respectively. The experiments showed that the size of double metal oxide nanosheets had an obvious effect on the water flux and dye rejection percentage of the microfiltration filter paper (Figure 5.37d–f). Double metal oxide nanosheets prepared by a 12-hour hydrothermal treatment possessed smaller sizes and could fill into the pores formed by ultralong hydroxyapatite nanowires and cellulose fibers, resulting in decreased water flux of the microfiltration filter paper. However, double metal oxide nanosheets prepared by a 72-hour hydrothermal treatment had larger sizes and could rarely fill into the pores of the filter paper, but instead stacked against each other to block the water transporting channels, and decreased the water flux. In the case of double metal oxide nanosheets prepared by a 24-hour hydrothermal treatment, some nanosheets could partially fill into the pores of the filter paper to enhance the adsorption performance, and many nanosheets could distribute between the layers to make the structure more porous and thus increase the water flux, and the rejection percentage for the dye increased (Figure 5.37e and f). Although some reported separation membranes could efficiently separate Congo red, their water fluxes were usually low (lesser than 250 L m^{-2} h^{-1} bar^{-1}). Therefore, the addition of double metal oxide nanosheets is an effective strategy for regulating the porous structure of filtration membranes, which can achieve simultaneous enhancement of both separation efficiency and water flux.[81]

(3) Nanofiltration Filter Paper Based on Ultralong Hydroxyapatite Nanowires and Cellulose Fibers/Nanofibers

A filter paper with a large number of nanopores and channels is important for highly efficient nanofiltration to significantly increase the water flux without obviously sacrificing the separation efficiency. The author's research group developed a new kind of nanofiltration filter paper consisting of a cellulose nanofiber barrier layer on the ultralong hydroxyapatite nanowire/cellulose fiber substrate.[82] The porous network formed by interwoven ultralong hydroxyapatite nanowires and cellulose fibers could provide many pores and channels for rapid water transportation. The experiments showed that the pure water flux of the microfiltration filter paper consisting of ultralong hydroxyapatite nanowires and cellulose fibers was 544.4 L m^{-2} h^{-1} bar^{-1}, which was ~50 times that of the nanofiltration filter paper consisting of ultralong hydroxyapatite nanowires and cellulose nanofibers. Compared with the nanofiltration filter paper based on ultralong hydroxyapatite nanowires and cellulose nanofibers, the pure water flux of the nanofiltration filter paper consisting of ultralong hydroxyapatite nanowires, cellulose fibers, and cellulose nanofibers could be enhanced by 20% with rejection percentages higher than 95% for four kinds of dyes. Moreover, the nanofiltration filter paper consisting of ultralong hydroxyapatite nanowires, cellulose fibers, and cellulose nanofibers possessed a relatively a high rejection percentage for sodium sulfate (75.7%) and sodium chloride (65.8%).

The experiments indicated that both cellulose fibers and cellulose nanofibers could enhance the tensile strength of the inorganic fire-resistant paper consisting of only ultralong hydroxyapatite nanowires (1.96 MPa). The tensile strength of the filter paper sheets composed of ultralong hydroxyapatite nanowires/cellulose nanofibers, and ultralong hydroxyapatite nanowires/cellulose fibers was 15.63 MPa and 5.56 MPa, respectively, at a cellulose weight ratio of 20 wt.%, which were about 8 and 2.8 times that of the pure ultralong hydroxyapatite nanowire paper. The tensile strength of the filter paper composed of 50 wt.% ultralong hydroxyapatite nanowires and 50 wt.% cellulose nanofibers was very high (>50 MPa).

Although cellulose nanofibers could greatly enhance the mechanical strength of the filter paper, a loose structure was needed for the filter paper to provide enough pores and channels to achieve high water flux. As shown in Figure 5.38a, both cellulose fibers and cellulose nanofibers could decrease the bulk of the ultralong hydroxyapatite nanowire filter paper while increasing its tensile strength, and the cellulose nanofiber/ultralong hydroxyapatite nanowire filter paper had a lower bulk compared with that of the cellulose fiber/ultralong hydroxyapatite nanowire filter paper. Because cellulose nanofibers possessed smaller diameters and lengths, a tighter structure formed and more water channels were blocked when cellulose nanofibers filled in between ultralong hydroxyapatite nanowires. Cellulose fibers with micrometer-sized dimensions could form larger pores and channels in the filter paper. Furthermore, cellulose nanofibers had many polar groups on the surface; thus, the ultralong hydroxyapatite nanowire/cellulose nanofiber filter paper exhibited a good hydrophilicity, and its contact angle was ~45°. Cellulose fibers and ultralong hydroxyapatite nanowires exhibited good hydrophilicity owing to their abundant

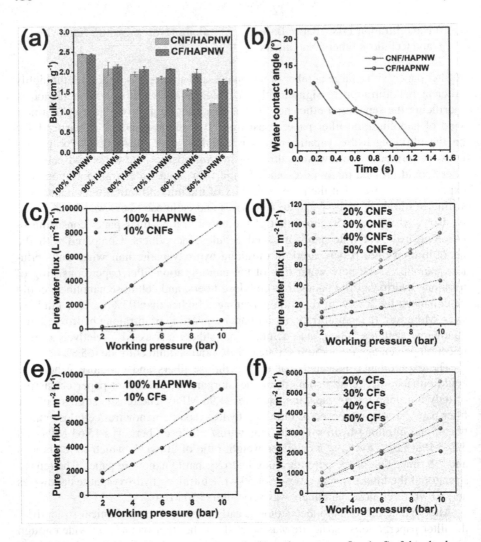

FIGURE 5.38 The bulk (a), water contact angle (b), and pure water flux (c–f) of the ultralong hydroxyapatite nanowire/cellulose fiber filter paper, and the ultralong hydroxyapatite nanowire/cellulose nanofiber filter paper sheets with different cellulose weight ratios. (Reprinted with permission from reference [82])

surface hydroxyl groups, and the contact angles of the ultralong hydroxyapatite nanowire/cellulose fiber filter paper were lower than those of the ultralong hydroxyapatite nanowire/cellulose nanofiber filter paper, indicating the cellulose fiber-based filter paper had a better hydrophilicity compared with the cellulose nanofiber-based filter paper (Figure 5.38b). Because the ultralong hydroxyapatite nanowire/cellulose fiber filter paper possessed a looser porous structure and better hydrophilicity, its pure water flux was significantly higher than that of the ultralong hydroxyapatite nanowire/cellulose nanofiber filter paper (Figure 5.38c–f). The pure water flux of the filter paper consisting of only ultralong hydroxyapatite nanowires was as high as 894.7 L m^{-2} h^{-1} bar^{-1}. However, the pure water flux of the ultralong hydroxyapatite nanowire/10 wt.% cellulose nanofiber filter paper reduced to 70.8 L m^{-2} h^{-1} bar^{-1} (Figure 5.38c). The pure water flux of the ultralong hydroxyapatite nanowire/cellulose nanofiber filter paper decreased from 10.8 L m^{-2} h^{-1} bar^{-1} to 3.0 L m^{-2} h^{-1} bar^{-1} when the cellulose weight ratio increased from 20 wt.% to 50 wt.% (Figure 5.38d). The pure water flux of the ultralong hydroxyapatite nanowire/10 wt.% cellulose fiber filter paper was 640.4 L m^{-2} h^{-1} bar^{-1} (Figure 5.38e). Furthermore, when the weight ratio of cellulose fibers increased from 20 wt.% to 50 wt.%, the pure water flux decreased from 544.4 L m^{-2} h^{-1} bar^{-1} to 201.4 L m^{-2} h^{-1} bar^{-1} (Figure 5.38f). The pure water flux of the ultralong hydroxyapatite nanowire/cellulose fiber filter paper was ~50 times that of the ultralong hydroxyapatite nanowire/cellulose nanofiber filter paper at a cellulose weight ratio of 20 wt.%.[82]

The bubble point method was used to measure the interconnected pores of the filter paper. The pore size distribution curves of two kinds of filter paper were obviously different because of different sizes of cellulose fibers and cellulose nanofibers. For the ultralong hydroxyapatite nanowire/cellulose nanofiber filter paper, cellulose nanofibers with small diameters of 5–10 nm could only fill among ultralong hydroxyapatite nanowires, forming smaller pores. The average pore size of the ultralong hydroxyapatite nanowire/cellulose nanofiber filter paper decreased from 105.2 nm to 62.6 nm when the cellulose weight ratio increased from 20 wt.% to 40 wt.%. For the ultralong hydroxyapatite nanowire/cellulose fiber filter paper, large-sized cellulose fibers acted as the skeleton of the filter paper, and ultralong hydroxyapatite nanowires could fill in the macropores formed by cellulose fibers. The average pore size increased from 527.6 nm to 993.0 nm when the cellulose weight ratio increased from 20 wt.% to 40 wt.%. The ultralong hydroxyapatite nanowire/cellulose fiber filter paper possessed looser porous structure, better hydrophilicity, higher water flux, and larger average pore size, and could be used as an ideal porous substrate for the construction of high-performance nanofiltration filter paper.

The filtration performance of the nanofiltration filter paper consisting of ultralong hydroxyapatite nanowires, cellulose fibers, and cellulose nanofibers was investigated using aqueous solutions containing different dyes. As shown in Figure 5.39a, the nanofiltration filter paper possessed a pure water flux of 10.9 L m^{-2} h^{-1} under a working pressure of 4 bar, which was ~20% higher than that of the nanofiltration filter paper consisting of ultralong hydroxyapatite nanowires and cellulose nanofibers. In addition, rejection percentages of the nanofiltration filter paper consisting of ultralong hydroxyapatite nanowires, cellulose fibers, and cellulose nanofibers under 4 bar working pressure for 500 ppm methyl blue, Congo red, orange G, and methyl orange were higher than 95% (Figure 5.39b). The cross-sectional schematic diagram of the nanofiltration filter paper consisting of ultralong hydroxyapatite nanowires, cellulose fibers, and cellulose nanofibers is shown in Figure 5.39c. The upper layer was a cellulose nanofiber barrier layer, and the lower porous substrate was an ultralong hydroxyapatite nanowire/cellulose fiber filter paper. Because the surface of 2,2,6,6-tetramethylpiperidine-1-oxyl free radical-oxidized cellulose fibers possessed a large number of carboxyl groups, they exhibited a strong Donnan effect on the

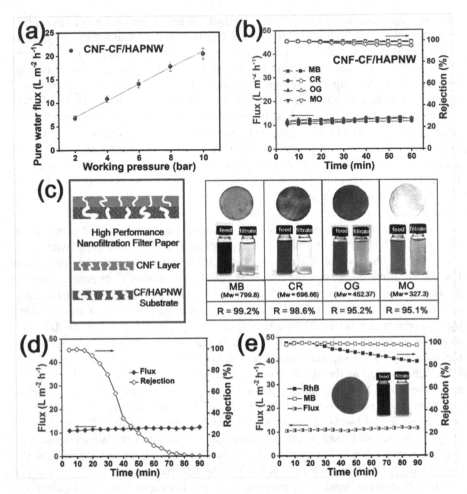

FIGURE 5.39 Dye filtration performance of the nanofiltration filter paper consisting of ultralong hydroxyapatite nanowires, cellulose fibers, and cellulose nanofibers. (a) Pure water flux versus working pressure; (b) water flux and rejection percentage for different dyes versus time; (c) schematic illustration of the nanofiltration filter paper (left), digital images of the nanofiltration filter paper after filtration, original and filtrate solutions, molecular weights, and rejection percentages of dyes; (d) water fluxes and rejection percentages of rhodamine B (Rh B) aqueous solution (500 ppm) by the nanofiltration filter paper under a working pressure of 4 bar; (e) water fluxes and rejection percentages of rhodamine B and methyl blue mixed aqueous solution (500 ppm each) using the nanofiltration filter paper under a working pressure of 4 bar (The weight ratio of cellulose fibers in all substrates was 20 wt.%). Abbreviation: methyl blue (MB), Congo red (CR), methyl orange (MO), orange G (OG), rhodamine B (Rh B). (Reprinted with permission from reference [82])

dyes with a negative charge. Moreover, the rejection percentage for the four dyes decreased with decreasing molecular weight of the dye, revealing that the separation performance of the nanofiltration filter paper for the dye was dependent on the size of dye molecule (Figure 5.39c). The experiments indicated that rejection percentages of the nanofiltration filter paper consisting of ultralong hydroxyapatite nanowires, cellulose fibers, and cellulose nanofibers for four types of dyes were only slightly lower than those of the nanofiltration filter paper consisting of ultralong hydroxyapatite nanowires and cellulose nanofibers (Figure 5.39c), but the water flux of the nanofiltration filter paper consisting of ultralong hydroxyapatite nanowires, cellulose fibers and cellulose nanofibers increased by 20%. The nanofiltration filter paper consisting of ultralong hydroxyapatite nanowires, cellulose fibers, and cellulose nanofibers could effectively adsorb rhodamine B, and the adsorption for 500 ppm rhodamine B aqueous solution was saturated in 90 minutes at a pressure of 4 bar. Using an aqueous solution containing 500 ppm of methyl blue and 500 ppm of rhodamine B, the rejection percentage of the nanofiltration filter paper consisting of ultralong hydroxyapatite nanowires, cellulose fibers, and cellulose nanofibers for methyl blue and rhodamine B within 20 minutes was close to 100% (Figure 5.39d). After 90 minutes of filtration, the rejection percentages for methyl blue and rhodamine B were 97.8% and 83.4%, respectively (Figure 5.39e). A similar phenomenon was also found for the nanofiltration filter paper consisting of ultralong hydroxyapatite nanowires and cellulose nanofibers. The adsorption quantities of the nanofiltration filter paper consisting of ultralong hydroxyapatite nanowires, cellulose fibers, and cellulose nanofibers for methyl blue, Congo red, orange G, and rhodamine B were 57.6 mg g^{-1}, 58.2 mg g^{-1}, 28.0 mg g^{-1}, 3.7 mg g^{-1}, and 93.0 mg g^{-1}, respectively. Obviously, the nanofiltration filter paper consisting of ultralong hydroxyapatite nanowires, cellulose fibers, and cellulose nanofibers exhibited a small adsorption capacity for negatively charged dyes because the nanofiltration filter paper itself was negatively charged, implying that the nanofiltration filter paper mainly depended on the Donnan exclusion for the removal of negatively charged dyes. Furthermore, the removal of the dye by the nanofiltration filter paper consisting of ultralong hydroxyapatite nanowires, cellulose fibers, and cellulose nanofibers was stable for a relatively long period of time.[82]

The molecular weight cut-off and rejection percentage of the nanofiltration filter paper consisting of ultralong hydroxyapatite nanowires, cellulose fibers, and cellulose nanofibers for monovalent and divalent ions were measured. The molecular weight cut-off of the nanofiltration filter paper was calculated by filtering electrically neutral poly(ethylene glycol) molecules. Usually, when the rejection percentage is 90%, the corresponding molecular weight is defined as the molecular weight cut-off of the filter paper. As shown in Figure 5.40a, the molecular weight cut-off of the nanofiltration filter paper was ~18.6 kDa, and thus the calculated Stokes radius of poly(ethylene glycol) with a molecular weight of 18.6 kDa was 4.0 nm, indicating that the average pore diameter of the nanofiltration filter paper was ~4.0 nm. For the nanofiltration filter paper, the rejection percentages for monovalent ions and divalent ions are key indicators of its filtration performance. The rejection percentages of the nanofiltration filter paper were measured for four kinds of salts (Na_2SO_4, NaCl, $MgSO_4$, and $MgCl_2$) at a concentration of 500 ppm under a working pressure

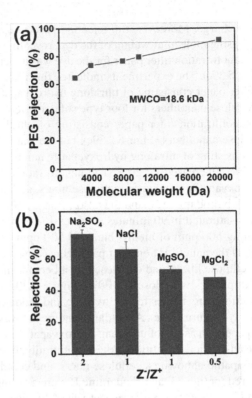

FIGURE 5.40   (a) Molecular weight cut-off (MWCO) and (b) rejection percentages of the nanofiltration filter paper consisting of ultralong hydroxyapatite nanowires, cellulose fiber and cellulose nanofibers for monovalent and divalent ions. (Reprinted with permission from reference [82])

of 4 bar. The experiments indicated that the rejection percentages for salt solutions were in the following descending order: Na_2SO_4 (75.7%) > NaCl (65.8%) > $MgSO_4$ (53.2%) > $MgCl_2$ (48.0%) (Figure 5.40b).[82]

As discussed in Section 4.7, the layer-structured catalytic fire-resistant paper consisting of ultralong hydroxyapatite nanowires and gold nanoparticles could be used for the catalytic conversion or degradation of organic pollutants in wastewater. The catalytic performance of the layer-structured catalytic fire-resistant paper in continuous flow reactions at room temperature was investigated.[64] The reduction of relatively toxic nitro-aromatic compounds to form corresponding amino-aromatic compounds and the degradation of organic dyes are important for wastewater treatment. The catalytic activity of the catalytic fire-resistant paper was evaluated by employing a model reduction reaction of 4-nitrophenol to 4-aminophenol by sodium borohydride in aqueous solution. A digital image of the testing device is shown in Figure 5.41a. An aqueous solution containing 4-nitrophenol and sodium borohydride simply flowed through the catalytic fire-resistant paper. The original solution containing 4-nitrophenol and sodium borohydride exhibited a yellow color and a strong characteristic absorption peak of 4-nitrophenol at ~400 nm (Figure 5.41b), while the

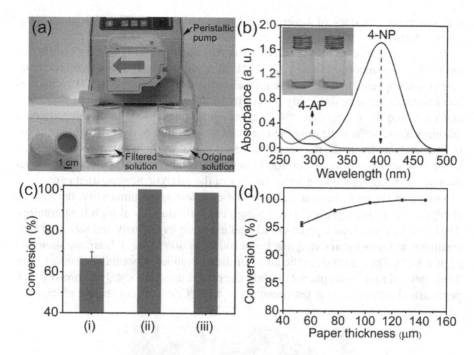

FIGURE 5.41 The catalytic performance of the layered catalytic fire-resistant paper for the continuous flow catalytic reduction reaction of 4-nitrophenol to 4-aminophenol in an aqueous solution containing 4-nitrophenol and sodium borohydride. (a) Digital images of the testing equipment and a filter container loaded with the catalytic fire-resistant paper (inset); (b) UV–visible absorption spectra and the digital image of an aqueous solution (inset) containing 4-nitrophenol and sodium borohydride before (light yellow color) and after (colorless) flowing through the catalytic fire-resistant paper; (c) comparison of conversion efficiencies of 4-nitrophenol to 4-aminophenol using three kinds of the catalytic fire-resistant paper sheets with different contents of gold nanoparticles; (d) the effect of the thickness of the catalytic fire-resistant paper on the conversion efficiency of 4-nitrophenol to 4-aminophenol. (Reprinted with permission from reference [64])

filtered solution flowing through the catalytic fire-resistant paper was colorless, indicating efficient catalytic reduction of 4-nitrophenol to 4-aminophenol. The intensity of the characteristic absorption peak of 4-nitrophenol decreased and that of 4-aminophenol peak increased in the UV–visible spectrum during the catalytic process, and the conversion efficiency of 4-nitrophenol to 4-aminophenol was 100% (Figure 5.41b). A control experiment was conducted, in which the original solution containing 4-nitrophenol and sodium borohydride was filtered through the fire-resistant paper without loading gold nanoparticles, and no catalytic activity was found. In addition, it was found that the content of gold nanoparticles had an effect on the catalytic performance of the catalytic fire-resistant paper (Figure 5.41c). The thickness of the catalytic fire-resistant paper could also affect the conversion efficiency of 4-nitrophenol. With increasing paper thickness from 53 μm to 103 μm and to 127 μm, the conversion efficiency of 4-nitrophenol increased from 95.6% to 99.5% to 100% (Figure 5.41d).

In addition to high catalytic activity, the recyclability and long-term stability of the catalyst are important factors for practical applications. The flow catalytic reaction of 4-nitrophenol to 4-aminophenol was continuously repeated for 20 cycles, and no obvious decrease in the catalytic activity of the catalytic fire-resistant paper was observed, and the conversion efficiency of 4-nitrophenol to 4-aminophenol could maintain above 99%, indicating the excellent recyclability of the catalytic fire-resistant paper. In addition, the flow catalytic reaction of 4-nitrophenol to 4-aminophenol was continuously performed for 200 hours, and a conversion efficiency was higher than 90%, indicating the long-term stability of the catalytic fire-resistant paper.

Owing to the high thermal stability and excellent nonflammability, the layered catalytic fire-resistant paper showed a high catalytic stability at high temperatures. The catalytic fire-resistant paper could well maintain its integrity and size after heat treatment at temperatures ranging from 100°C to 400°C for 1 hour, as shown in Figure 5.42a. The conversion efficiency for the continuous flow catalytic reduction of 4-nitrophenol to 4-aminophenol slightly decreased using the catalytic fire-resistant paper after heat treatment at temperatures of >300°C, but the conversion efficiencies

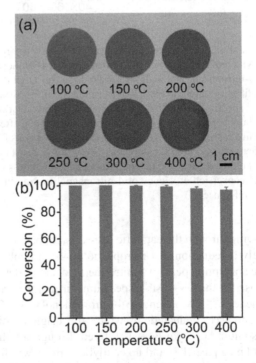

FIGURE 5.42 (a) Digital images and (b) conversion efficiencies for the continuous flow catalytic reduction of 4-nitrophenol to 4-aminophenol using the layered catalytic fire-resistant paper sheets after heat treatment at different temperatures for 1 hour. (Reprinted with permission from reference [64])

were still higher than 96%, showing the high catalytic stability of the catalytic fire-resistant paper under high-temperature conditions (Figure 5.42b). Merits such as high catalytic activity, high thermal stability, and excellent nonflammability of the layer-structured catalytic fire-resistant paper have promising applications in various catalytic areas such as high-temperature catalysis, catalytic purification of automobile exhaust gas, and high-temperature industrial waste gas.

The catalytic activities of the layer-structured catalytic fire-resistant paper for continuous flow catalytic degradation of organic dyes in aqueous solution in the presence of sodium borohydride were investigated. Figure 5.43a shows the outlet concentration of rhodamine B in the filtered solution when an aqueous solution

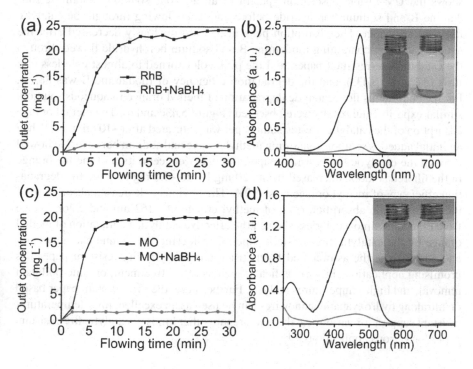

FIGURE 5.43 Properties of the layer-structured catalytic fire-resistant paper for continuous flow catalytic degradation of organic dyes in aqueous solution. (a) The outlet concentrations of rhodamine B (RhB) in the filtered solution when the aqueous solution containing rhodamine B and sodium borohydride (or only rhodamine B without sodium borohydride) flowed through the catalytic fire-resistant paper; (b) UV–visible absorption spectra and the digital image (inset) of an aqueous solution containing rhodamine B and sodium borohydride before (pink color) and after (almost colorless) flowing through the catalytic fire-resistant paper; (c) the outlet concentrations of methyl orange (MO) in the filtered solution when the aqueous solution containing methyl orange and sodium borohydride (or only methyl orange without sodium borohydride) flowed through the catalytic fire-resistant paper; (d) UV–visible absorption spectra and the digital image (inset) of an aqueous solution containing methyl orange and sodium borohydride before (orange color) and after (almost colorless) flowing through the catalytic fire-resistant paper. (Reprinted with permission from reference [64])

containing rhodamine B and sodium borohydride (or only rhodamine B) flowed through the catalytic fire-resistant paper. The outlet rhodamine B concentration increased with increasing flow time of the aqueous solution containing only rhodamine B without sodium borohydride, and the value was nearly 25 mg L^{-1} at 30 minutes, indicating that rhodamine B adsorption of the catalytic fire-resistant paper was saturated after ~15 minutes. However, the outlet concentration of rhodamine B was at a low level (approximately 1.5 mg L^{-1}) when an initial aqueous solution containing rhodamine B and sodium borohydride flowed through the catalytic fire-resistant paper because of the effective degradation of rhodamine B by the catalytic effect of the catalytic fire-resistant paper in addition to the adsorption effect. Figure 5.43b shows the UV–visible absorption spectra of an aqueous solution containing rhodamine B and sodium borohydride before and after flowing through the catalytic fire-resistant paper. The absorption peak at 552 nm obviously decreased after the aqueous solution containing rhodamine B and sodium borohydride flowed through the catalytic fire-resistant paper, and the pink color turned to almost colorless (the inset of Figure 5.43b), and the degradation efficiency of rhodamine B was 88.1%. For the continuous flow catalytic degradation of methyl orange in aqueous solution, similar experimental results were observed (Figure 5.43c and d). The methyl orange adsorption of the catalytic fire-resistant paper was saturated after ~10 minutes. When an initial aqueous solution containing methyl orange and sodium borohydride flowed through the catalytic fire-resistant paper, the outlet concentration of methyl orange in the filtered solution decreased from ~20 mg L^{-1} to ~1.8 mg L^{-1}, and the degradation efficiency of methyl orange was ~90%. The obviously decreased absorbance of the characteristic absorption peak of methyl orange at ~462 nm and color change from orange to almost colorless showed the effective catalytic degradation of methyl orange by the catalytic fire-resistant paper. Considering the unique characteristics and advantages, the as-prepared layer-structured catalytic fire-resistant paper has promising applications in various fields, such as water treatment, organic pollutant removal, and high-temperature catalysis. Furthermore, the fire-resistant paper based on ultralong hydroxyapatite nanowires can be used as an excellent high-temperature-resistant support for various catalysts to prepare other kinds of catalytic fire-resistant paper for various applications.[64]

5.5.3. Solar Energy–Driven Desalination of Seawater

Some traditional desalination technologies need to directly or indirectly consume non-renewable fossil fuel resources, and their disadvantage is that it will cause environmental pollution and other problems. Therefore, choosing renewable energy for desalination of seawater is an effective way to solve the crisis of freshwater shortage. Solar energy is an efficient and sustainable source of clean energy. Using solar energy to drive seawater desalination is very promising for the practical application. Solar energy was investigated for the desalination of seawater; however, the water evaporation efficiency and output of clean water were usually low. This was mainly due to the low sunlight absorption efficiency of water, and as most of the heat is absorbed by the water body. In order to improve the evaporation efficiency of water in the solar-driven desalination of seawater, on the one hand, it is necessary to use light-to-heat conversion materials to increase the absorption efficiency of sunlight and the efficiency of light-to-heat conversion; on the other hand, according to the feature that water evaporation only occurs on the surface of the water body, a self-floating material is needed to collect and focus the heat generated by the photo-thermal conversion material at the air–water interface to reduce heat loss. In recent years, researchers studied a variety of photothermal conversion materials used to generate water vapor and clean water under solar light irradiation. However, many photothermal conversion materials have disadvantages such as poor thermal stability and low efficiency.

As discussed in Section 4.8, the author's research group developed a highly flex-ible photothermal fire-resistant inorganic paper comprising a fire-resistant paper made from ultralong hydroxyapatite nanowires and a small amount of glass fibers as the thermal insulation support and a layer of carbon nanotubes as the light absorber for solar energy–driven seawater desalination and water purification.[65] The highly porous networked structure, excellent high-temperature stability, and fire-resistance merits of the photothermal fire-resistant paper are desirable for solar energy–driven water evaporation. The thermal conductivity of the fire-resistant paper without car-bon nanotubes was 0.085 W m^{-1} K^{-1}, and that of the photothermal fire-resistant paper was 0.088 W m^{-1} K^{-1}, which were significantly lower than that of water (0.6 W m^{-1} K^{-1}). The excellent thermal insulation property of the photothermal fire-resistant paper could effectively minimize the heat loss from the paper surface to bulk water, which was desirable for realizing the high-performance solar energy–driven water evaporation.[65]

The light absorption efficiency and photothermal properties of the as-prepared photothermal fire-resistant paper were investigated. As shown in Figure 5.44a, the photothermal fire-resistant paper exhibited a high absorption efficiency (~96%) in the visible and infrared regions. However, the fire-resistant paper without carbon nanotubes had a low absorption efficiency of ~5%. The high optical absorption effi-ciency of the photothermal fire-resistant paper could be attributed to the high light absorption ability of carbon nanotubes, and the porous and rough upper paper sur-face, which is favorable for achieving high photothermal conversion performance. As shown in Figure 5.44b, the surface temperature of the photothermal fire-resistant paper rapidly increased to ~80°C and then was stable under 1 kW m^{-2} solar light irradiation. In comparison, the fire-resistant paper without carbon nanotubes exhib-ited a low surface temperature of ~35°C. The maximum surface temperature of the photothermal fire-resistant paper was ~80°C, 95°C, 120°C, and 145°C at a solar light power intensity of 1 kW m^{-2}, 2 kW m^{-2}, 3 kW m^{-2}, and 4 kW m^{-2}, respectively (Figure 5.44c). The infrared thermal images indicated that the surface temperature distribution on the photothermal fire-resistant paper was uniform under solar light irradiation (Figure 5.44b), implying that carbon nanotubes were uniformly distrib-uted on the surface of the photothermal fire-resistant paper. Figure 5.44d shows infrared thermal images of water without or in the presence of the photothermal fire-resistant paper under 1 kW m^{-2} solar light irradiation for different times. The tem-perature distribution of the bulk water without or in the presence of the photothermal

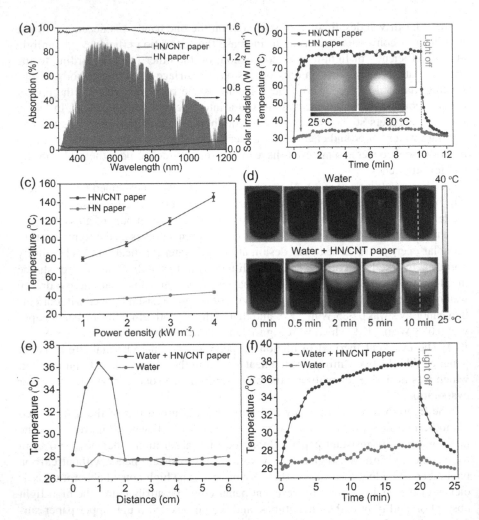

FIGURE 5.44 (a) Solar light absorption spectra of the photothermal fire-resistant (HN/CNT) paper consisting of ultralong hydroxyapatite nanowires and carbon nanotubes, and the fire-resistant (HN) paper without carbon nanotubes; (b) surface temperatures of the photothermal fire-resistant paper, and the fire-resistant paper without carbon nanotubes under 1 kW m⁻² solar light irradiation in air for different times, and insets are the corresponding infrared thermal images of the paper sheets under solar light irradiation for 10 minutes; (c) equilibrium maximum surface temperatures of the photothermal fire-resistant paper, and the fire-resistant paper without carbon nanotubes under solar light irradiation at different power densities in air for 10 minutes; (d) infrared thermal images of water without or in the presence of the photothermal fire-resistant paper under 1 kW m⁻² solar light irradiation for 0 minute, 0.5 minute, 2 minutes, 5 minutes, and 10 minutes, respectively; (e) temperature distributions of the marked line from top to bottom in (d) after 10 minutes solar light irradiation; (f) surface temperatures of water or the photothermal fire-resistant paper floating on the water surface under 1 kW m⁻² solar light irradiation for different times. (Reprinted with permission from reference [65])

fire-resistant paper was uniform (~26°C) before solar light irradiation. Under 1 kW m^{-2} solar light irradiation for 10 minutes, the temperature of the water surface without the photothermal fire-resistant paper was ~27°C. In comparison, the surface temperature of the photothermal fire-resistant paper was 37°C after 10 minutes solar light irradiation. Furthermore, an obvious temperature gradient was observed in the presence of the photothermal fire-resistant paper (Figure 5.44e), indicating the heat localization effect of the photothermal fire-resistant paper. The solar light irradiation time had an effect on the surface temperature. As shown in Figure 5.44f, the surface temperature of pure water increased slowly to ~29°C after 20 minutes solar light irradiation. However, the surface temperature of the photothermal fire-resistant paper increased to ~38°C after solar light irradiation for 20 minutes.

The solar energy–driven water evaporation performance of the photothermal fire-resistant paper floating on the water surface was quantitatively studied. As shown in Figure 5.45a, the water weight decreased approximately linearly with solar light irradiation time, and the water weight loss in the presence of the photothermal fire-resistant paper was 1.09 kg m^{-2} after 1 hour of solar light irradiation at a power density of 1 kW m^{-2}, which was much higher than that of pure water in dark (0.11 kg m^{-2}) or pure water under solar light irradiation (0.30 kg m^{-2}) or other control samples. The curves of water evaporation rate versus solar light irradiation time under different solar light power densities are shown in Figure 5.45b. The water evaporation rate increased with increasing solar light irradiation time and power density, and the water evaporation rate was 1.09 kg m^{-2} h^{-1}, 3.38 kg m^{-2} h^{-1}, 5.71 kg m^{-2} h^{-1}, 8.09 kg m^{-2} h^{-1}, and 11.87 kg m^{-2} h^{-1} under 1 kW m^{-2}, 3 kW m^{-2}, 5 kW m^{-2}, 7 kW m^{-2}, and 10 kW m^{-2} solar light irradiation, respectively. However, the water evaporation rate of pure water in the absence of the photothermal fire-resistant paper was much smaller (0.30 kg m^{-2} h^{-1}, 0.58 kg m^{-2} h^{-1}, 0.81 kg m^{-2} h^{-1}, 1.09 kg m^{-2} h^{-1}, and 1.56 kg m^{-2} h^{-1}) under the same conditions. The enhancement factor is defined as the ratio of water evaporation rate of the photothermal fire-resistant paper to that of the blank sample (pure water without the photothermal fire-resistant paper). The enhancement factor had a minimum value of 3.63 under 1 kW m^{-1} solar light irradiation, and 7.61 under 10 kW m^{-2} solar light irradiation (Figure 5.45c).

The photothermal fire-resistant paper exhibited higher water evaporation efficiency than pure water without the photothermal fire-resistant paper and other control samples. The water evaporation efficiency in the presence of the photothermal fire-resistant paper was 70.3%, 73.6%, 75.7%, 79.0%, and 83.1% under 1 kW m^{-2}, 3 kW m^{-2}, 5 kW m^{-2}, 7 kW m^{-2}, and 10 kW m^{-2} solar light irradiation, respectively (Figure 5.45d). Furthermore, in order to further enhance the water evaporation rate and evaporation efficiency, a device comprising a polystyrene foam and a piece of cotton fiber paper was designed and prepared, and it could minimize the heat transfer from the photothermal fire-resistant paper to the bulk water and ambient atmosphere. By using this device, the water evaporation rate and water evaporation efficiency of the photothermal fire-resistant paper increased to 1.31 kg m^{-2} h^{-1} and 83.2% under 1 kW m^{-2} solar light irradiation, and to 14.31 kg m^{-2} h^{-1} and 92.8% under 10 kW m^{-2} solar light irradiation, respectively.[65]

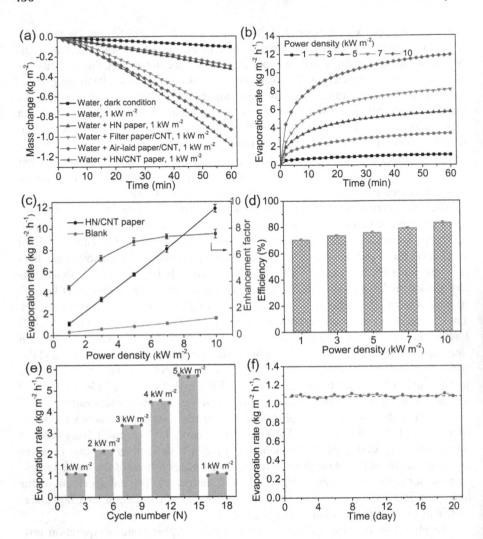

FIGURE 5.45 Solar energy–driven water evaporation properties of the photothermal fire-resistant paper. (a) Cumulative water weight loss versus solar light irradiation time under various conditions; (b) water evaporation rate in the presence of the photothermal fire-resistant paper versus solar light irradiation time at different solar light power densities; (c) water evaporation rate in the presence of the photothermal fire-resistant paper and the control group without the photothermal fire-resistant paper at different solar light power densities, and the enhancement factor of the photothermal fire-resistant paper compared with the control group without the photothermal fire-resistant paper at different solar light power densities; (d) efficiencies for solar energy–driven water evaporation in the presence of the photothermal fire-resistant paper at different solar light power densities; (e) water evaporation rate versus cycle number at different solar light power densities in the presence of the photothermal fire-resistant paper; (f) the long-time water evaporation stability of the photothermal fire-resistant paper in 20 days under solar light irradiation at a power density of 1 kW m^{-2} for 1 hour each day. (Reprinted with permission from reference [65])

The recycling and long-time performances of the photothermal fire-resistant paper for solar energy–driven water evaporation were also investigated. As shown in Figure 5.45e, the water evaporation performance was stable during 18 solar light irradiation cyclic tests at a series of solar light power densities, exhibiting the excellent stability and recycling ability of the photothermal fire-resistant paper. In addition, the photothermal fire-resistant paper showed excellent performance in long-time water evaporation, as shown in Figure 5.45f.

The photothermal fire-resistant paper can be used for the generation of fresh drinkable water from seawater and wastewater. A portable device was designed and prepared for solar energy–driven water evaporation using the photothermal fire-resistant paper, as shown in Figure 5.46a. Several kinds of seawater samples, including simulated seawater and genuine seawater, were investigated. As shown in Figure 5.46b, the concentrations of sodium chloride in the generated condensed water obtained from five simulated seawater samples after solar photothermal desalination using the photothermal fire-resistant paper decreased by three orders of magnitude compared with those of the original solutions, and the ion rejection percentages were >99.95%. The collected clean water using the photothermal fire-resistant paper could meet the standards of healthy drinkable water as defined by the World Health Organization and U. S. Environmental Protection Agency. The real seawater obtained from the East China Sea was also investigated, and the experimental results are shown in Figure 5.46c, indicating that the concentrations of five primary ions (Na^+, K^+, Mg^{2+}, Ca^{2+}, and B^{3+}) significantly decreased with ion rejection percentages >94.32% after solar photothermal desalination using the photothermal fire-resistant paper, and the concentrations of ions were far below the ion concentrations for drinkable water as defined by the World Health Organization. These experimental results demonstrated the excellent performance of the as-prepared photothermal fire-resistant paper in seawater desalination.

The photothermal fire-resistant paper could also be used for solar photothermal purification of wastewater containing heavy metal ions. As shown in Figure 5.46d, the concentrations of heavy metal ions such as Ni^{2+}, Cu^{2+}, Cd^{2+}, Mn^{2+} were high in the original wastewater samples, but the concentrations of metal ions in the collected purified water after solar photothermal purification using the photothermal fire-resistant paper were < 2 µg L^{-1} with metal ion rejection percentages >99.99%, which were lower than the concentrations as defined by the drinking water standard (GB5749-2006).

Furthermore, the photothermal fire-resistant paper could also be used for solar photothermal purification of wastewater containing organic dyes and bacteria. After solar photothermal purification using the photothermal fire-resistant paper, the purified water from aqueous solutions containing methyl orange and rhodamine B was colorless and transparent (Figure 5.46e), and the UV–visible absorption absorbance value was near zero, indicating the high quality of the obtained clean water. The photothermal fire-resistant paper exhibited a high performance in purifying wastewater containing bacteria such as Gram-negative E. coli under solar light irradiation, and the purified water was transparent and colorless, and no bacteria colony was observed on the solid nutrient agar plate (Figure 5.46f). The quality of the purified

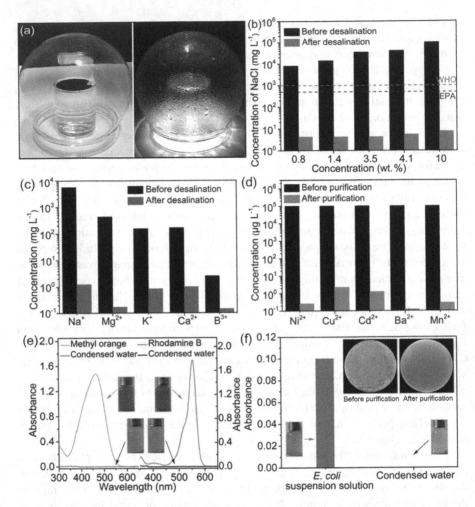

FIGURE 5.46 Performances of the photothermal fire-resistant paper in solar energy–driven seawater desalination and wastewater purification. (a) Digital images of a portable lab-made device for solar energy–driven photothermal water evaporation and clean water collection; (b) concentrations of sodium chloride in five kinds of simulated seawater samples (0.8 wt.%, the Baltic Sea; 1.4 wt.%, the North Sea; 3.5 wt.%, the average world sea; 4.1 wt.%, the Red Sea; and 10 wt.%, the Dead Sea) and the obtained clean water after solar photothermal desalination; the dashed lines in (b) are the standards for the healthy drinkable water of the World Health Organization (WHO) and U. S. Environmental Protection Agency (EPA); (c) concentrations of five primary ions in the genuine seawater obtained from the East China Sea and the obtained clean water after solar photothermal desalination; (d) concentrations of heavy metal ions in the simulated wastewater and the purified water after solar photothermal purification; (e) UV–visible absorption spectra and the corresponding digital images of aqueous solutions of methyl orange, rhodamine B, and the obtained water after solar energy–driven photothermal purification; (f) UV–visible absorbance values and the corresponding digital images of the aqueous solution containing *E. coli* and the condensed water after solar photothermal purification, digital images in the upper right corner in (f) show the solid nutrient agar plates after culture with *E. coli* suspension (left) and the condensed water after solar photothermal purification (right). (Reprinted with permission from reference [65])

water after solar photothermal purification using the photothermal fire-resistant paper could meet the national drinking water standard (GB5749-2006). Compared with other technologies for water purification, such as reverse osmosis, electrodialysis, and nanofiltration, the solar energy–driven water purification using the photothermal fire-resistant paper is more convenient, highly efficient, and low cost, and is promising for application in the production of the clean drinkable water from seawater and wastewater.[65]

One of the problems is that the local salt concentration continuously increases during the seawater desalination process, leading to the accumulation of salt crystals on the photothermal paper, which can impair light absorption, hinder water transportation and vapor escape, and deteriorate water evaporation performance. The salt fouling can be mitigated by appropriate interfacial designs. For example, using a hydrophobic surface can prevent the infiltration of salt aqueous solutions, and salt ions can be blocked underneath the hydrophobic layer and further be made to diffuse to the bulk solution through hydrophilic channels, resulting in salt rejection and clean top surface.[83] To solve the salt accumulation problem, the author's research group developed a new kind of the flexible salt-rejecting photothermal fire-resistant paper comprising reduced graphene oxide, ultralong hydroxyapatite nanowires, and glass fibers for high-performance solar energy–driven water evaporation and stable seawater desalination without salt accumulation.[83] The salt-rejecting photothermal fire-resistant paper possessed merits such as an hierarchical porous structure, interconnected channels, high mechanical strength, high efficiencies of solar light absorption and photothermal conversion, fast water transportation, good heat insulation, and salt-rejecting ability, which could efficiently produce clean water from seawater and wastewater with high rejection percentages of organic dyes, metal ions, and salt ions.

The water evaporation properties of the salt-rejecting photothermal fire-resistant paper and control samples are shown in Figure 5.47. Figure 5.47a shows the time-dependent weight changes of pure water without or in the presence of different paper samples. The maximum water evaporation rate of the salt-rejecting photothermal fire-resistant paper was $1.48 \, kg \, m^{-2} \, h^{-1}$, which was 5.10 times that of pure water in the absence of the salt-rejecting photothermal fire-resistant paper under 1 sun illumination (1 kW m^{-2}). The water evaporation efficiency of the salt-rejecting photothermal fire-resistant paper was as high as 89.2% under 1 sun illumination (Figure 5.47b), which was comparable to or even higher than many photothermal materials reported in the literature (Figure 5.47c). The long-time water evaporation performance of the salt-rejecting photothermal fire-resistant paper is shown in Figure 5.47d, indicating that the water evaporation rate of the salt-rejecting photothermal fire-resistant paper was relatively stable with time and excellent recycling performance.

To further evaluate the practical performance of the salt-rejecting photothermal fire-resistant paper under natural sunlight illumination, a solar energy–driven water evaporation device including water evaporation, vapor condensation, and clean water collection was designed and prepared. The device consisted of a source water container, a floating foam covered with an air-laid paper for water transportation, a circular salt-rejecting photothermal fire-resistant paper on the air-laid paper, and a vapor

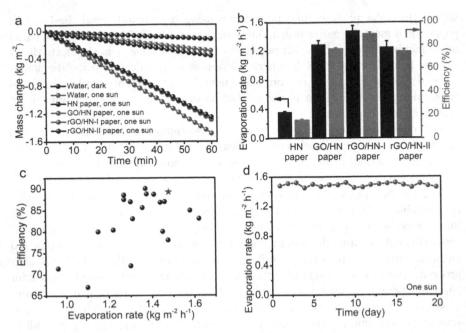

FIGURE 5.47 Water evaporation performances of the salt-rejecting photothermal fire-resistant paper and control samples under 1 sun illumination (1 kW m⁻²). (a) Weight change of pure water versus solar light irradiation time; (b) water evaporation rates and efficiencies of different paper samples; (c) comparison of water evaporation rates and efficiencies of the salt-rejecting photothermal fire-resistant paper (labeled with a pentagram) and other solar evaporators listed in reference [83]; (d) water evaporation rates of the salt-rejecting photothermal fire-resistant paper within 20 days, and the solar light irradiation time was 1 hour each day. HN paper stands for the fire-resistant paper consisting of only ultralong hydroxyapatite nanowires; GO/HN paper stands for the photothermal fire-resistant paper consisting of ultralong hydroxyapatite nanowires and graphene oxide without heat treatment; rGO/HN-I stands for the hydrophilic photothermal fire-resistant paper without the salt-rejecting ability consisting of ultralong hydroxyapatite nanowires and reduced graphene oxide after heat treatment at 150°C for 2 hours in vacuum; and rGO/HN-II stands for the hydrophobic salt-rejecting photothermal fire-resistant paper consisting of ultralong hydroxyapatite nanowires and reduced graphene oxide after heat treatment at 150°C for 6 hours in vacuum. (Reprinted with permission from reference [83])

condensation chamber, as shown in Figure 5.48a. Under an outdoor natural sunlight illumination in Shanghai, China on July 22, 2019, clean water was collected from the bottom of the container (Figure 5.48b). After 8 hours of continuous operation, the clean water production amount was 6.06 kg m⁻². The practical solar intensity, ambient temperature, and collected purified water volume during the testing process are shown in Figure 5.48c. The wastewater purification performance of the salt-rejecting photothermal fire-resistant paper was tested, and the industrial wastewater sample from a chemical factory was used without any treatment. As shown in Figure 5.48d, the original wastewater sample exhibited a dark red color and a high conductivity

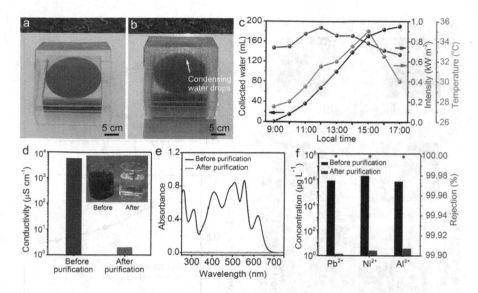

FIGURE 5.48 Digital images and water purification performances of the solar energy–driven water purification device consisting of the salt-rejecting photothermal fire-resistant paper. (a, b) Digital images of the solar energy–driven water purification device; (c) changes in solar light intensity, ambient temperature, and volume of collected clean water under the natural sunlight illumination from 9:00 am to 17:00 pm on July 22, 2019, in Shanghai, China; (d) conductivity values of the wastewater sample before and after purification and the corresponding digital images; (e) UV–visible absorption spectra of the wastewater sample before and after purification; (f) metal ion concentrations of the wastewater sample before and after purification, and rejection percentages. The industrial wastewater sample was obtained from a chemical factory without any treatment. (Reprinted with permission from reference [83])

of ~5700 μS cm^{-1}; however, the collected clean water after purification was colorless and transparent with a small conductivity of 2.0 μS cm^{-1}, which could meet the healthy drinkable water standard. Figure 5.48e shows the UV–visible absorption spectra of the original wastewater sample and purified water, indicating that the pollutants were effectively removed after the solar photothermal purification process. Furthermore, metal ion concentrations were measured before and after the solar photothermal water purification process. The concentrations of Pb^{2+}, Ni^{2+}, and Al^{3+} ions in the purified water were 1.39 μg L^{-1}, 2.51 μg L^{-1}, and 3.68 μg L^{-1}, respectively, which were lower than concentration limits of World Health Organization, and the ion rejection percentages of the salt-rejecting photothermal fire-resistant paper were higher than 99.99% (Figure 5.48f). In addition, the decontamination properties of the salt-rejecting photothermal fire-resistant paper for organic dyes such as Congo red, rose bengale, brilliant green, and other metal ions Fe^{3+}, Cu^{2+}, and Ni^{2+} in the simulated wastewater were also investigated, and the experimental results showed that the removal efficiencies for these pollutants were close to 100%.[83]

In addition, the salt-rejecting performances of the hydrophilic photothermal fire-resistant (rGO/HN-I) paper and hydrophobic salt-rejecting photothermal

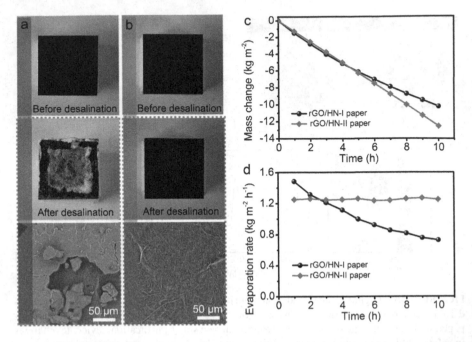

FIGURE 5.49 Solar energy–driven desalination performance of the photothermal fire-resistant paper. (a) Digital images and SEM image of the hydrophilic photothermal fire-resistant paper (rGO/HN-I) before and after the desalination; (b) digital images and SEM image of the hydrophobic salt-rejecting photothermal fire-resistant paper (rGO/HN-II) before and after the desalination; (c) weight change and (d) evaporation rate of an aqueous solution containing 3.5 wt.% sodium chloride under 1 kW m^{-2} solar light illumination in the presence of the hydrophilic photothermal fire-resistant paper and hydrophobic salt-rejecting photothermal fire-resistant paper. (Reprinted with permission from reference [83])

fire-resistant (rGO/HN-II) paper were studied via the solar energy–driven desalination of an aqueous solution containing 3.5 wt.% sodium chloride as the simulated seawater. After continuous solar energy–driven desalination for 10 hours, a white layer of salt crystals was observed on the black hydrophilic photothermal fire-resistant paper, as shown in Figure 5.49a. The SEM image indicated that deposited salt crystals had sizes ranging from several to tens of micrometers. In addition, the weight change of the simulated seawater using the hydrophilic photothermal fire-resistant paper was not linear with solar light irradiation time (Figure 5.49c). The water evaporation rate gradually decreased with increasing solar light irradiation time (Figure 5.49d) because of the formation of salt crystals on the paper surface. Salt accumulation could decrease light absorption ability, block the pores and channels for water transportation and vapor escape, and deteriorate long-time desalination performance. Intermittent operation, daily cleaning, and rinsing with water are the possible solutions to solve the salt accumulation problem, but these methods are time-consuming and increase the cost, which limit their practical applications.

Fortunately, the experimental results indicated that the surface of the hydrophobic salt-rejecting photothermal fire-resistant paper was clean after 10 hours of continuous solar photothermal desalination, and the SEM image showed the salt crystal-free paper surface (Figure 5.49b). The weight change of the simulated seawater using the hydrophobic salt-rejecting photothermal fire-resistant paper exhibited an approximately linear relationship with solar light irradiation time (Figure 5.49c). Although the water evaporation rate of the hydrophobic salt-rejecting photothermal fire-resistant paper was lower at the early stage, after three hours it was higher than that of the hydrophilic photothermal fire-resistant paper, and it was steady throughout the entire desalination process, exhibiting a high performance in continuous solar photothermal desalination (Figure 5.49d).

Furthermore, the long-time solar energy–driven desalination of the hydrophobic salt-rejecting photothermal fire-resistant paper was studied using the actual seawater, and the experimental results are shown in Figure 5.50. The water evaporation rate was highly stable for 20 days of continuous solar energy–driven seawater desalination, and the surface of the hydrophobic salt-rejecting photothermal fire-resistant paper was clean and salt-free, indicating its long-term durability and excellent recycling performance (Figure 5.50a). In addition, the quality of purified water produced from the actual seawater was excellent (Figure 5.50b), in contrast to the high concentrations of Na^+, K^+, Mg^{2+}, Ca^{2+}, and B^{3+} ions in the actual seawater sample, the ion concentrations in purified water obtained after solar photothermal desalination significantly decreased to <4 mg L^{-1}. The rejection percentages of Na^+, K^+, Mg^{2+}, and Ca^{2+} ions were >99.0%, and the rejection percentage of B^{3+} ions was >95.0% after solar energy–driven photothermal desalination (Figure 5.50c). The ion concentrations in the purified water obtained after solar energy–driven seawater desalination in the presence of the hydrophobic salt-rejecting photothermal fire-resistant paper could meet the drinking water standards of the World Health Organization and US Environmental Protection Agency, indicating high purification performance of the hydrophobic salt-rejecting photothermal fire-resistant paper for the production of clean drinkable water.[83]

5.5.4. FILTER PAPER FOR POLLUTED AIR PURIFICATION AND ANTI-HAZE FACE MASK

In recent years, the haze weather frequently appeared, and severe air pollution has aroused people's concerns for health. As we know, $PM_{2.5}$ fine particles in air can severely threaten people's health. $PM_{2.5}$ refers to the particulate matter with an ambient aerodynamic diameter ≤ 2.5 μm. The higher the content of $PM_{2.5}$ in air, the more serious the air pollution. Under normal circumstances, larger particles with particle sizes larger than 10 μm will be blocked outside the human nose; particles with particle sizes between 2.5 μm and 10 μm will be partially blocked by the villi of the nose, although some can enter the upper respiratory tract but can be excreted through the sputum, etc., which are relatively less harmful to human health. However, the physiological structure of the human body does not have the ability to effectively filter and block $PM_{2.5}$ fine particles. $PM_{2.5}$ fine particles have small particle sizes, high activity,

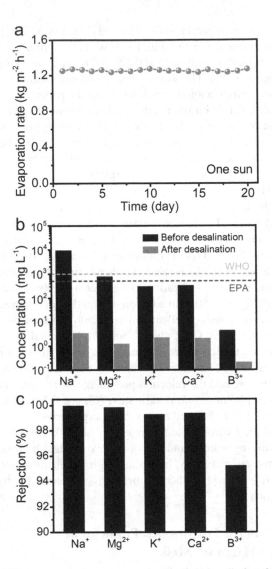

FIGURE 5.50 (a) Water evaporation rate versus solar light irradiation time at a solar power density of 1 kW m⁻² for 20 days (8 hours light irradiation each day) using the hydrophobic salt-rejecting photothermal fire-resistant paper and the real seawater obtained from the East China Sea; (b) ion concentrations and (c) rejection percentages of Na^+, K^+, Mg^{2+}, Ca^{2+}, and B^{3+} ions in real seawater obtained from the East China Sea before and after solar energy–driven desalination. (Reprinted with permission from reference [83])

and long residence time in the atmosphere, long transporting distance, and wide range of negative effects. After inhalation into the human body, $PM_{2.5}$ fine particles will enter the bronchus and even lungs, causing a variety of diseases. In addition, $PM_{2.5}$ fine particles are also carriers of toxic and harmful substances, viruses, and bacteria to fuel the spread of diseases. Therefore, it is of great significance to develop high-efficiency filter materials for air $PM_{2.5}$ fine particles.

The most common air filtration product used in daily life is the face mask. Common gauze, nonwoven fabrics, and other filter materials are commonly used in face masks, and the fibers are relatively thick and the pores between the fibers are large. The common face masks usually have a high efficiency for removing large particles with sizes larger than 10 μm (PM_{10}) but difficult to efficiently remove $PM_{2.5}$ fine particles in air. A major problem that currently exists is that many face masks on the market have low filtration efficiency for $PM_{2.5}$ fine particles. In addition to the unsatisfactory filtering effect of $PM_{2.5}$ fine particles, some face masks have poor air permeability. If the inhalation resistance is too large, people may experience dizziness and chest tightness after wearing face masks. Therefore, there is an urgent need to develop high-performance filter materials and face masks for protection from polluted air with high filtration efficiency and good air permeability to meet the needs of people's daily life and work.

The author's research group developed a new kind of air filter paper consisting of ultralong hydroxyapatite nanowires and cotton fibers with high removal efficiencies for $PM_{2.5}$ and PM_{10},[84] as shown in Figure 5.51a–c. Ultralong hydroxyapatite nanowires could interweave with each other and with cotton fibers to form the highly porous air filter paper. By optimizing the weight ratio of ultralong hydroxyapatite nanowires to cotton fibers, the as-prepared air filter paper exhibited high removal efficiencies of >95% for $PM_{2.5}$ and PM_{10}, a lower pressure drop, and a smaller thickness than that in a commercial face mask. As a proof-of-concept, a homemade anti-haze face mask was prepared by imbedding the as-prepared air filter paper into a commercial

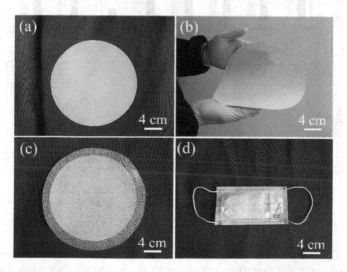

FIGURE 5.51 (a–c) Digital images of a new kind of air filter paper consisting of ultralong hydroxyapatite nanowires and cotton fibers with high removal efficiencies for $PM_{2.5}$ and PM_{10}; (d) digital image of an anti-haze face mask prepared using the as-prepared air filter paper. (Reprinted with permission from reference [84])

face mask (Figure 5.51d). In addition, other functions such as antibacterial activity could be achieved by the immobilization of silver nanoparticles as the representative bactericide, enabling the preparation of the multifunctional air filter paper.

The $PM_{2.5}$ and PM_{10} used in the experiments were prepared by burning incense, considering that the resultant smoke possessed similar ingredients as polluted air during a hazy day. The generated smoke particles showed a broad size distribution

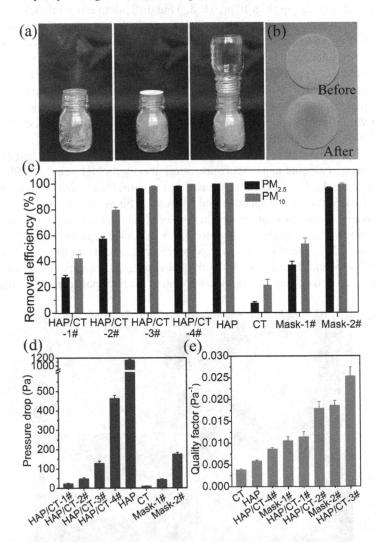

FIGURE 5.52 The filtration performance of the as-prepared air filter paper consisting of ultralong hydroxyapatite nanowires and cotton fibers. (a) The air filter paper was used for blocking the diffusion of smoke from the bottom bottle to the outer space or the top bottle; (b) digital images of the air filter paper before and after the filtration test; (c–e) quantitative results of air filtration tests for different samples: (c) the removal efficiencies for $PM_{2.5}$ and PM_{10}; (d) pressure drop; (e) quality factor. (Reprinted with permission from reference [84])

with a large proportion of particulate matter with small sizes (e.g., 0.3 μm and 0.5 μm). A simple setup was designed for the demonstration of the removal of the particulate matter by the as-prepared air filter paper. As shown in Figure 5.52a, the as-prepared air filter paper could effectively block the diffusion of the smoke from the bottom bottle to the outer space or to the top bottle. After some time, the top bottle was still clear and the concentrations of $PM_{2.5}$ and PM_{10} in the top bottle were below 35 μg m^{-3}. The digital images in Figure 5.52b show an obvious color change of the air filter paper from white to light yellow before and after the test, revealing that the air filter paper had an excellent performance in air filtration.

The removal efficiencies for $PM_{2.5}$ and PM_{10}, and the pressure drop of a series of air filter paper sheets and two kinds of commercial face masks were studied. The air filter paper sheets with hydroxyapatite/cotton weight ratios of 1:4, 2:3, 3:2, and 4:1 were denoted as HAP/CT-1, HAP/CT-2, HAP/CT-3, and HAP/CT-4, respectively. The air filter paper with a higher hydroxyapatite/cotton weight ratio showed a better air filtration performance. The removal efficiencies were higher than 95% for both $PM_{2.5}$ and PM_{10} when the hydroxyapatite/cotton weight ratio was 3:2 or higher (Figure 5.52c). The enhanced removal efficiencies for $PM_{2.5}$ and PM_{10} originated from the dense, highly porous networked structure, high specific surface area, and abundant mesopores formed by ultralong hydroxyapatite nanowires.

Unfortunately, the value of the pressure drop increased with increasing hydroxyapatite/cotton weight ratio (Figure 5.52d). There was an optimal hydroxyapatite/cotton weight ratio to achieve a good balance between the removal efficiency for $PM_{2.5}$ and the pressure drop. The air filter paper with a hydroxyapatite/cotton weight ratio of 3:2 was selected for further study, and it showed a high removal efficiency of 96.08% and 97.55% for $PM_{2.5}$ and PM_{10}, respectively, and a pressure drop of 128 Pa. These removal efficiencies were 8.5 and 4.6 times that of the pure cotton fiber paper (11.34% and 21.04% for $PM_{2.5}$ and PM_{10}, respectively) with the same basis weight of 30 g m^{-2}, indicating that the integration of ultralong hydroxyapatite nanowires and cotton fibers could significantly increase the removal efficiencies of $PM_{2.5}$ and PM_{10}.

The removal efficiency of the as-prepared air filter paper was higher than that of a commercial face mask with a similar pressure drop. For example, the air filter paper (HAP/CT-2) exhibited the removal efficiencies of 57.65% and 79.71% for $PM_{2.5}$ and PM_{10}, which were 1.6 and 1.5 times, respectively, that of a commercial face mask-1 with a similar pressure drop. Moreover, the air filter paper showed a lower pressure drop than a commercial face mask with similar removal efficiencies for $PM_{2.5}$ and PM_{10}. As an example, the air filter paper (HAP/CT-3) had a pressure drop of 128 Pa, which was lower than 176 Pa of the commercial face mask-2 with similar removal efficiencies for $PM_{2.5}$ and PM_{10}. Another advantage was that the air filter paper was significantly thinner than the commercial face mask. The thickness of the air filter paper (HAP/CT-3) was only ~110 μm, while the commercial breathing mask-2 had a thickness of ~1.10 mm. The larger thickness could lead to a higher pressure drop. In addition, the overall filtration performance of the filter materials can be evaluated by the quality factor considering both the removal efficiency and pressure drop. The optimal air filter paper exhibited a higher quality factor than commercial face masks (Figure 5.52e).[84]

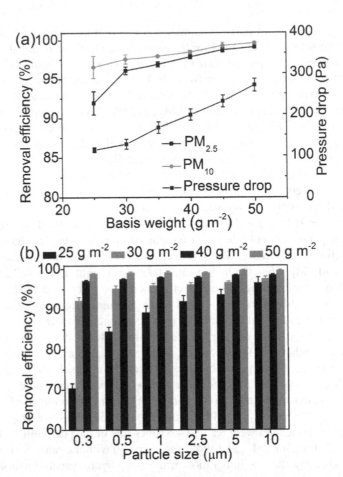

FIGURE 5.53 Particulate matter removal efficiency of the as-prepared air filter paper versus basis weight or particle size. (a) Effects of the basis weight on the removal efficiency and the pressure drop; (b) removal efficiencies for various particulate matter particle sizes. (Reprinted with permission from reference [84])

In addition to the hydroxyapatite/cotton weight ratio, another critical parameter affecting the air filtration performance was the basis weight of the air filter paper. As shown in Figure 5.53, both the removal efficiency and the pressure drop increased with increasing basis weight of the air filter paper. The increased basis weight of the air filter paper could increase the contact points for capturing particulate matter particles; however, the extended tortuous air-flow channels could result in enhanced air resistance simultaneously. Moreover, the increased basis weight of the air filter paper affected the removal efficiency of large particles (e.g. 2.5 μm and 10 μm) relatively slightly but that of small particles (e.g. 0.3 μm) was affected greatly. For example, the PM_{10} removal efficiency increased from 96.53% to 97.55% with increasing basis weight from 25 g m^{-2} to 30 g m^{-2}, and the $PM_{2.5}$ removal efficiency increased from 91.94% to 96.08%. For small particles of $PM_{0.3}$, the removal efficiency was greatly increased from 70.40% to 92.19% to 97.02% when the basis weight increased from 25 g m^{-2} to 30 g m^{-2} to 40 g m^{-2}, respectively. During the filtration process, large particles (e.g. 2.5 μm and 10 μm) were captured by ultralong hydroxyapatite nanowires through both interception and inertial impaction, while small particles (e.g. 0.3 mm) were captured primarily through diffusion effect.

The effects of the air-flow velocity on the removal efficiency and pressure drop of the as-prepared air filter paper were investigated. As shown in Figure 5.54a, the removal efficiencies of PM$_{2.5}$ and PM$_{10}$ decreased with increasing air-flow velocity, but the removal efficiency of PM$_{0.3}$ decreased more greatly. On the other hand, the pressure drop increased almost linearly with increasing air-flow velocity. For example, the pressure drop increased from 82 Pa to 225 Pa when the air-flow velocity increased from 3 cm s^{-1} to 8 cm s^{-1} (Figure 5.54a). In addition, the removal efficiencies for PM$_{2.5}$ of the air filter paper at various initial PM$_{2.5}$ concentrations were

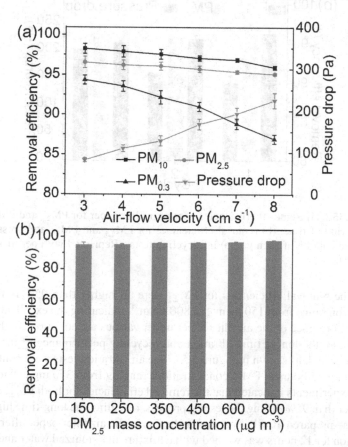

FIGURE 5.54 Particulate matter removal efficiency of the as-prepared air filter paper versus air-flow velocity or initial PM$_{2.5}$ mass concentration. (a) Effect of air-flow velocity on the removal efficiency and pressure drop; (b) effect of initial PM$_{2.5}$ mass concentration on the removal efficiency. (Reprinted with permission from reference [84])

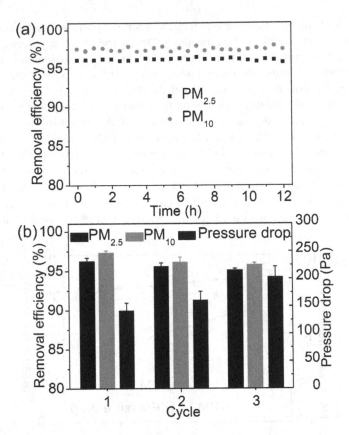

FIGURE 5.55 (a) Removal efficiencies of the air filter paper for $PM_{2.5}$ and PM_{10} during a longer period of time; (b) removal efficiencies for $PM_{2.5}$ and PM_{10}, and pressure drop values (third column of each group) in recycling tests. (Reprinted with permission from reference [84])

studied. The removal efficiencies for $PM_{2.5}$ were all higher than 95% in the $PM_{2.5}$ concentration range from 150 μg m^{-3} to 800 μg m^{-3}, indicating the excellent air purification performance of the air filter paper under various air pollution conditions.[84]

In addition, the longer time filtration and recycling performance of the air filter paper were tested, as shown in Figure 5.55. The air filtration tests were continuously performed for 12 hours at $PM_{2.5}$ concentrations ranging from 350 mg m^{-3} to 400 mg m^{-3}. The experiments indicated that the removal efficiencies for both $PM_{2.5}$ and PM_{10} were higher than 95% during the testing process, revealing the long-time high stability of the as-prepared air filter paper. Furthermore, the air filter paper after continuous filtration for 12 hours was washed with a mixture of deionized water and ethanol, and the obtained recycled clean suspension of ultralong hydroxyapatite nanowires and cotton fibers was adopted for the preparation of a new air filter paper. After three cycles, the removal efficiencies for $PM_{2.5}$ and PM_{10} were still higher than 95%. However, the pressure drop increased obviously (Figure 5.55b). This phenomenon could be explained by the residual particulate matter particles adsorbed on ultralong hydroxyapatite nanowires and cotton fibers, resulting in the increased air resistance of the air filter paper.

In order to enhance the mechanical properties of the air filter paper, a poromeric and medical gauze with ~1 mm pore size was used as the supporting layer. As a proof-of-concept, a gauze-supported air filter paper was used and embedded into a commercial face mask for personal protection from the polluted air. The high removal efficiencies (>95%) for both $PM_{2.5}$ and PM_{10}, and relatively low air resistance (~160 Pa) could be achieved, implying the potential practical application of the as-prepared air filter paper.

Because of the unique physicochemical properties of ultralong hydroxyapatite nanowires, various chemical compounds and nanoparticles can be immobilized on the surface of ultralong hydroxyapatite nanowires, enabling the preparation of the nanocomposites with multifunctional properties. The inhalable microorganisms can be disseminated through the diffusion of particulate matter particles, which are responsible for various human diseases. Furthermore, the air filters can be easily contaminated by microorganisms due to the aggregation of organic compounds on the filters after long-time use. Therefore, it is highly desirable to develop air filters with antibacterial properties. In order to address this issue, silver nanoparticles as the representative bactericidal agent were immobilized on the surface of ultralong hydroxyapatite nanowires, and the antibacterial air filter paper was prepared. The agar diffusion method was adopted to determine the antibacterial activity against the *E. coli* (*E. coli*) and *S. aureus* bacteria. The experiments indicated that there was no inhibition zone for the air filter paper without silver nanoparticles; however, the air filter paper with silver nanoparticles could produce a circular inhibition zone for both *E. coli* and *S. aureus*, revealing its high antibacterial activity. The simple integration of silver nanoparticles with ultralong hydroxyapatite nanowires and the resulting excellent antibacterial properties demonstrated that the as-prepared air filter paper is a promising material for loading multifunctional agents to develop various novel kinds of air filters. The new kind of air filter paper may be produced continuously using a paper making machine. Given the unique merits, the air filter paper based on ultralong hydroxyapatite nanowires is promising for applications in the purification of polluted air and personal protection from polluted air.[84]

5.6. ENERGY-RELATED APPLICATIONS

The fire-resistant paper based on ultralong hydroxyapatite nanowires has unique merits such as excellent resistance to both high temperatures and fire, high flexibility, nanoscale porous networked structure, and adjustable hydrophilicity and hydrophobicity, surface functional groups for surface modification, etc. In the energy field, high-temperature and fire-resistant materials are highly desirable for many applications. It is expected that the fire-resistant paper based on ultralong hydroxyapatite nanowires is promising for various applications in the energy field. Recently, the energy-related applications of the fire-resistant paper based on ultralong hydroxyapatite nanowires were investigated, such as the high-temperature-resistant separator for advanced lithium-ion battery, high-temperature-resistant electrode material for advanced lithium-ion battery, and high-temperature-resistant separator and electrode material for high-energy-density lithium-ion capacitor. These research works will be discussed in the following sections.

5.6.1. HIGH-TEMPERATURE-RESISTANT SEPARATOR FOR ADVANCED LITHIUM-ION BATTERY

Lithium-ion batteries have the advantages of high energy density, high power density, and long cycle life, so they are widely used in portable electronic devices such as notebook computers, mobile phones, digital cameras, and other electric products. In recent years, with the rapid development of new clean energy vehicles, higher requirements have been placed on the performance and safety of new power batteries and energy-storage batteries. New battery materials will provide new opportunities and possibilities for the improvement of the battery performance and safety.

Lithium-ion batteries are mainly composed of five parts: positive electrode material, negative electrode material, separator, electrolyte, and packaging materials. Among them, the battery separator mainly plays the role of preventing direct contact between the positive and negative electrodes, preventing battery short circuit, and diffusing ions. The battery separator is a key material to ensure the battery safety and it can affect the battery performance. Although the separator does not directly participate in the electrochemical reactions of the battery, its performance can affect the interfacial structure and internal resistance of the battery, which in turn influence the energy density, cycle life, and rate of the battery. In addition, the thermal stability of the separator also determines the temperature tolerance and the safety of the battery. The ideal battery separator should have superior properties, such as excellent electrical insulation, high mechanical strength, high electrochemical stability, high thermal stability, high porosity, suitable pore size, excellent electrolyte wettability, and high electrolyte adsorption capacity.

The current commercial lithium-ion battery separators are mainly organic polyolefin separators. The advantages of such organic separators are low price, good mechanical properties, and high electrochemical stability. However, the polyolefin separators also have shortcomings, for example: (1) the polyolefin battery separators have low polarity, but the electrolytes usually use organic solvents, so the polyolefin separators exhibit poor wettability with the electrolytes; (2) the polyolefin separators usually adopt a stretching method to prepare a porous structure, but the stretching method not only has high requirements for material processing equipment, but it is also difficult to obtain high porosity, resulting in large internal electric resistance of the battery; (3) polyolefin separators have poor thermal stability, and are easy to shrink or melt at relatively high temperatures, causing direct contact between the positive and negative electrodes of the battery, resulting in short circuit of the battery, fire, or even explosion accidents.

The author's research group developed a new kind of highly flexible and highly porous battery separator based on ultralong hydroxyapatite nanowires with high thermal stability, good fire resistance, excellent electrolyte wettability and a high absorption capacity, and superior electrochemical properties.[85] A hierarchical cross-linked network structure was formed between ultralong hydroxyapatite nanowires and cellulose fibers by hybridization, which enabled the separator to have high flexibility and good mechanical strength. The high thermal stability of ultralong hydroxyapatite nanowires enabled the separator to maintain its structural integrity at temperatures as high as 700°C, and the fire-resistant property of ultralong hydroxyapatite nanowires could ensure high safety of the battery. Benefiting from its unique composition and highly porous structure, the as-prepared high-temperature-resistant separator showed near zero contact angle with the liquid electrolyte and high electrolyte uptake of 253%, exhibiting excellent electrolyte wettability. The high-temperature-resistant separator is promising for application in advanced lithium-ion batteries and other kinds of batteries with enhanced performance and high safety.

Figure 5.56 shows digital images of the as-prepared high-temperature-resistant separator based on ultralong hydroxyapatite nanowires, showing high flexibility of the separator under different bending conditions. The high-temperature-resistant separator could be rolled, twisted, folded, scrunched, and unscrunched with no visible damages, indicating the good mechanical strength and high flexibility of the novel kind of high-temperature-resistant separator. The mechanical properties of the separator are very important for battery safety. An ideal separator should have high tensile strength to withstand the high tension upon assembling and casual collisions,

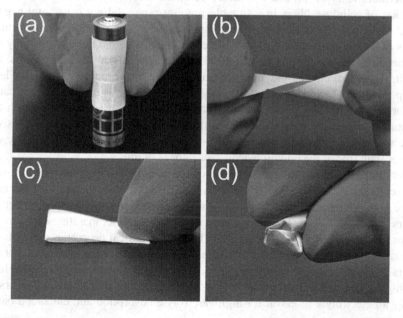

FIGURE 5.56 Digital images of the as-prepared high-temperature-resistant separator based on ultralong hydroxyapatite nanowires, showing high flexibility of the separator under different bending conditions: (a) rolled; (b) twisted; (c) folded; (d) scrunched. (Reprinted with permission from reference [85])

and prevent internal short circuits induced from debris of the rough electrode and dendrite growth.

The high porosity of the high-temperature-resistant separator was favorable for liquid electrolyte infiltration and adsorption, and thus could improve the cyclability and rate capability of batteries. Open, continuous, and interconnected nanopores of the high-temperature-resistant separator could form from the interwoven ultralong hydroxyapatite nanowire networks. The porosity of the as-prepared high-temperature-resistant separator consisting of ultralong hydroxyapatite nanowires and cellulose fibers was as high as ~81%, which was close to that of the pure ultralong hydroxyapatite nanowire high-temperature-resistant separator without cellulose fibers (79%). The average pore size of the high-temperature-resistant separator was ~120.9 nm, which was slightly larger than that of the pure ultralong hydroxyapatite nanowire high-temperature-resistant separator without cellulose fibers (117.4 nm). Both the pure ultralong hydroxyapatite nanowire high-temperature-resistant separator without cellulose fibers and the high-temperature-resistant separator with cellulose fibers exhibited a narrow pore size distribution ranging from ~110 nm to 130 nm, which was favorable for its application as the high-temperature-resistant separator in the lithium-ion battery.

Furthermore, the as-prepared high-temperature-resistant separator consisting of ultralong hydroxyapatite nanowires and cellulose fibers possessed a Gurley value of 83 s, which was close to that of the pure ultralong hydroxyapatite nanowire high-temperature-resistant separator without cellulose fibers (95 s), revealing that the addition of cellulose fibers in a suitable amount range did not cause any negative effect to the porosity of the high-temperature-resistant separator. In comparison, the pores of the polypropylene separator were inconsecutive, and there were many nonporous domains, mainly because of the formation process of stretching pores. As a result, those discrete, slit-like pore structures of the polypropylene separator led to low porosity (47%) and high Gurley value (278 s), which was unfavorable for battery performance. The thickness of the pure ultralong hydroxyapatite nanowire high-temperature-resistant separator without cellulose fibers and high-temperature-resistant separator with cellulose fibers was about 47 μm and 56 μm, respectively. They were thicker than that of the polypropylene separator (~28 μm).

As discussed above, the cross-section of the as-prepared high-temperature-resistant separator exhibited a well-aligned layered structure. The thickness of the layers of the high-temperature-resistant separator was about 5 μm and the spacing between layers was around 1 μm. The layered structure of the high-temperature-resistant separator has several benefits, for example: (1) the mechanical properties are enhanced by increasing the interface interactions between layers; (2) the interspace between layers can alleviate large extrusion under external stress, improving the flexibility of the high-temperature-resistant separator; and (3) the interspace between layers can act as an electrolyte reservoir for high electrolyte adsorption and enhanced battery performance.

The tensile strength of commercial polypropylene separator usually depends on the orientation. The polypropylene separator had a tensile strength of 98.39 MPa along the machine direction and only 9.28 MPa along the transversal direction.

In comparison, the tensile strength of the pure ultralong hydroxyapatite nanowire high-temperature-resistant separator without cellulose fibers was ~6.7 MPa, but the tensile strength of the high-temperature-resistant separator consisting of ultralong hydroxyapatite nanowires and cellulose fibers increased to 13.21 MPa owing to the synergistic combination of the van der Waals force and hydrogen bonding. In addition, the tensile strength of the high-temperature-resistant separator was essentially the same along different directions. The high-temperature-resistant separator exhibited a significantly higher Young's modulus (440 MPa) compared with the polypropylene separator (154 MPa). The high Young's modulus of the separator is desirable for preserving the structural integrity and inhibiting the rupture in case of an accidental crash, further improving battery safety. Compared with the polypropylene separator at the transversal direction, the high-temperature-resistant separator had a higher tensile strength and higher Young's modulus, implying that it is promising for application in high-safety batteries. Furthermore, the tensile strength of the as-prepared high-temperature-resistant separator was much higher than those of many previously reported inorganic separators.[85] One serious safety concern for the polyolefin separators is their mechanical stability at elevated temperatures. The tensile strength of the polypropylene separator decreased rapidly with increasing temperature, and the polypropylene separator shrank and its mechanical strength decreased to almost zero at 175°C. In comparison, the high-temperature-resistant separator showed a high thermal stability.[85]

Nonpolar polyolefin-based separators usually exhibit poor compatibility with the conventional polar liquid electrolytes due to their intrinsically hydrophobic nature. The polypropylene separator had a contact angle of 65° after dropping the liquid electrolyte on its surface for 90 seconds. In contrast, the electrolyte droplet penetrated into the ultralong hydroxyapatite nanowire-based high-temperature-resistant separators in 5 seconds. The electrolyte diffusion heights for the ultralong hydroxyapatite nanowire-based high-temperature-resistant separator without and with cellulose fibers were 18 mm and 16 mm, respectively, which were much higher than that of the polypropylene separator (2 mm). The experimental results showed that the ultralong hydroxyapatite nanowire-based separator possessed superior electrolyte wettability than the polypropylene separator. In addition, the electrolyte uptake percentages of the ultralong hydroxyapatite nanowire-based high-temperature-resistant separators with and without cellulose fibers were 253% and 248%, respectively, much higher than that of the polypropylene separator (76%). The superior electrolyte wettability and high electrolyte adsorption capacity of the ultralong hydroxyapatite nanowire-based separator can be explained by the combination of favorable affinity to the electrolyte, high porosity, and unique layered structure.[85]

Thermal stability is another important factor for batteries, especially for high power and high energy ones, which can determine the safety and operation temperatures of batteries. Thermogravimetric and differential scanning calorimetry measurements were taken to evaluate the thermal properties of the as-prepared ultralong hydroxyapatite nanowire-based high-temperature-resistant separators (Figure 5.57a and b). The polypropylene separator showed a broad endothermic peak, corresponding to the softening and melting of polypropylene, and it decomposed at temperatures

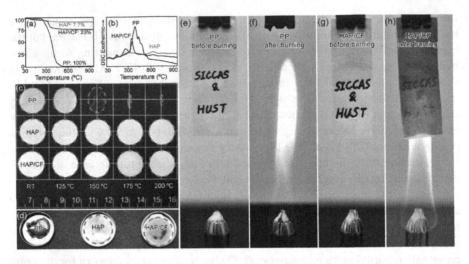

FIGURE 5.57 Thermal stability tests of the polypropylene (PP) separator, pure ultralong hydroxyapatite nanowire high-temperature-resistant separator without cellulose fibers (HAP), and ultralong hydroxyapatite nanowire high-temperature-resistant separator with cellulose fibers (HAP/CF). (a) Thermogravimetric curves and (b) differential scanning calorimetry curves of the separators; (c) thermal stability tests of the separators at different temperatures; (d) digital images of the separators after ignition, a liquid electrolyte (100 μL) was dropped on each separator before testing; (e, f) fire-resistant tests of the polypropylene separator before (e) and after (f) burning; (g, h) fire-resistant tests of the ultralong hydroxyapatite nanowire high-temperature-resistant separator with cellulose fibers before (g) and after (h) burning. (Reprinted with permission from reference [85])

higher than ~266°C, and was completely burned out at ~492°C. Owing to the high thermal stability of ultralong hydroxyapatite nanowires, no obvious peak appeared below 300°C in the differential scanning calorimetry curves of ultralong hydroxyapatite nanowire-based high-temperature-resistant separators with or without cellulose fibers. For the thermogravimetric curves, the weight loss for pure ultralong hydroxyapatite nanowire high-temperature-resistant separator was ~7.7%, resulting from the loss of adsorbed water and oleate groups. Even at a high temperature of 900°C, the ultralong hydroxyapatite nanowire-based high-temperature-resistant separators could still maintain 92.3% and 77% of the initial weights, respectively. Although cellulose fibers in the high-temperature-resistant separator would decompose at high temperatures, the skeletons made of ultralong hydroxyapatite nanowire networks could be well maintained. As shown in Figure 5.57c, the thermal shrinkage rate for the polypropylene separator was 18% at 125°C, and became more severe at high temperatures, for instance, more than 90% at 200°C. In comparison, the ultralong hydroxyapatite nanowire-based high-temperature-resistant separators showed nearly zero shrinkage from room temperature to 200°C. In particular, the pure ultralong hydroxyapatite nanowire high-temperature-resistant separator exhibited excellent thermal stability, and no change was observed both in color and dimension. Although the color of the high-temperature-resistant separator with cellulose fibers turned to light yellow because of the oxidation of cellulose fibers at 200°C, its shape and size could be well maintained. The high thermal stability of the ultralong hydroxyapatite nanowire-based high-temperature-resistant separators could greatly enhance the safety of the batteries when working at high temperatures, and significantly widen their applications in lithium-ion batteries and other kinds of batteries, especially in high-temperature environments.[85]

It is known that the currently used polyolefin separators and organic liquid electrolytes are extremely thermal sensitive and highly flammable. The ignition tests with flame were performed for the as-prepared ultralong hydroxyapatite nanowire-based high-temperature-resistant separators with the liquid electrolyte on their surface. When being exposed to fire, the high-temperature-resistant separators did not shrink throughout the whole test (Figure 5.57d). The electrolyte dropped on the pure ultralong hydroxyapatite nanowire high-temperature-resistant separator could not be ignited. Self-extinguishing occurred immediately even though the electrolyte on the high-temperature-resistant separator with cellulose fibers could be ignited, indicating its excellent flame-retardant property. In comparison, the polypropylene separator showed severe shrinkage, and the electrolyte dropped on its surface could be ignited and continuously combusted. The fire-resistant tests (Figure 5.57e–h) showed that the ultralong hydroxyapatite nanowire-based high-temperature-resistant separators were nonflammable, and their structural integrity could be well preserved on the flame of a spirit lamp. In contrast, the polypropylene separator immediately curled up, shrank severely, and was rapidly burned out in the flame.

Electrochemical stability measured by the linear-sweep voltammetry is one prerequisite for the separator. The onset potential of a steady increase in the characterized current density represents the electrochemical stability limitation of the electrolyte-soaked separator. The electrolyte-soaked high-temperature-resistant separator with cellulose fibers showed an anodic potential window up to 4.6 V versus Li^+/Li, higher than 4.4 V for the polypropylene separator. According to the Nyquist plots, the calculated ionic conductivity was 3.05 mS cm^{-1} for the high-temperature-resistant separator without cellulose fibers, and 3.07 mS cm^{-1} for the high-temperature-resistant separator with cellulose fibers, which was about seven times that of the polypropylene separator (0.43 mS cm^{-1}). It is well known that the ionic conductivity of a separator is greatly dependent on the electrolyte wettability, electrolyte uptake percentage, and porosity. High affinity and large uptake for the liquid electrolyte are favorable for fast ion transporting behavior of the separator. Furthermore, the unique layered structure and continuously interconnected nanopores of the high-temperature-resistant separator could decrease the ion diffusion barrier. The high ionic conductivity of the high-temperature-resistant separator was supported by the higher air permeability (lower Gurley value), lower McMullin number, and lower tortuosity. The lithium-ion transport number of the polypropylene separator, high-temperature-resistant separator without cellulose fibers, and high-temperature-resistant separator with cellulose fibers was 0.35, 0.54, and 0.52, respectively. The higher lithium-ion transport number of the separator based on ultralong hydroxyapatite nanowires indicated the higher Li$^+$ ion transport ability. The excellent liquid electrolyte affinity of the high-temperature-resistant separator could facilitate fast lithium-ion transporting. The abundant hydroxyl groups of ultralong hydroxyapatite nanowires and cellulose fibers could form hydrogen bonds with the fluorine atoms in PF_6^- anions of the electrolyte, which could hinder the movement of anions of the electrolyte and allow a higher fraction of Li^+ ions available for conduction.[85]

Figure 5.58 shows the electrochemical properties at room temperature of the cells prepared using different kinds of separators. The recycling performance and rate capability of the LiFePO$_4$/separator/Li half cells assembled using the

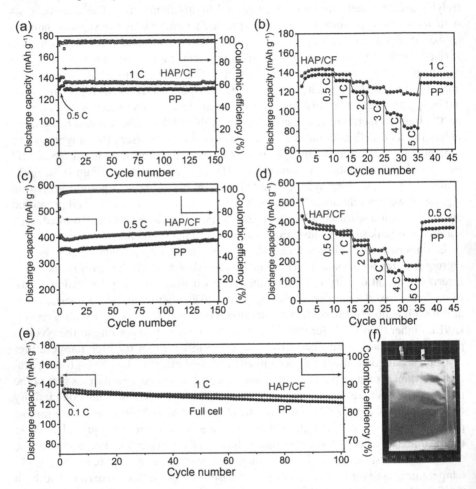

FIGURE 5.58 Electrochemical properties of the as-prepared cells using the ultralong hydroxyapatite nanowire high-temperature-resistant separator with cellulose fibers in comparison with the cells using the polypropylene separator at room temperature. (a) Recycling performance and (b) rate capability of the LiFePO$_4$/separator/Li half cells (1 C = 170 mAh g^{-1}); (c) recycling performance and (d) rate capability of the graphite/separator/Li half cells (1 C = 375 mAh g^{-1}); (e) recycling performance of the LiFePO$_4$/separator/graphite full cells (1 C = 140 mAh g^{-1}); (f) digital image of the as-assembled pouch-type full cell using the high-temperature-resistant separator. (Reprinted with permission from reference [85])

high-temperature-resistant separator with cellulose fibers and commercial poly-propylene separator are shown in Figure 5.58a and b. The cell with the high-tem-perature-resistant separator exhibited an initial discharge capacity of 138 mAh g^{-1}, which was higher than that using the polypropylene separator (130.1 mAh g^{-1}) at 0.5 C. During the recycling process, the cell with the high-temperature-resistant separator still showed a higher discharge capacity of 135.4 mAh g^{-1} than that of the polypropylene separator (129.5 mAh g^{-1}) after 145 cycles at 1 C, revealing superior cyclability of the high-temperature-resistant separator. The potential profiles showed typical electrochemical features of LiFePO$_4$, and the curve shape using the high-temperature-resistant separator was similar to that using the polypropylene separa-tor. Figure 5.58b shows the rate capability of the LiFePO$_4$/separator/Li half cells at different current densities. A capacity retention of 85.7% was obtained for the cell with the high-temperature-resistant separator at 5 C compared with its initial dis-charge capacity. In comparison, the capacity retention was only 65.7% for the cell with the polypropylene separator. In addition, the discharge capacity of the cell with the high-temperature-resistant separator was ~117.2 mAh g^{-1}, which was 1.4 times that of the cell with the polypropylene separator (83 mAh g^{-1}) at 5 C. The superior rate capability of the high-temperature-resistant separator could be explained by the higher ionic conductivity and lower interfacial resistance compared with the poly-propylene separator.

Furthermore, the cell with the high-temperature-resistant separator showed a bet-ter recycling performance and rate capability using the working electrode of graph-ite compared with the polypropylene separator (Figure 5.58c and d). The discharge capacity of the cell with the high-temperature-resistant separator was ~421 mAh g^{-1}, which was higher than that with the polypropylene separator (386 mAh g^{-1}) after 150 cycles at 0.5 C. As shown in Figure 5.58d, the cell with the high-temperature-resistant separator showed a much better rate capability (175 mAh g^{-1} at 5 C) than that with the polypropylene separator (101.5 mAh g^{-1}). The experiments indicated that no new redox peaks were observed in the cyclic voltammogram curves for the cell assembled with the high-temperature-resistant separator, indicating that the high-temperature-resistant separator had no obvious interference with the electro-chemical reactions on the LiFePO$_4$ cathode and graphite anode. The gaps between redox peaks of the cell with the high-temperature-resistant separator were narrower and the shapes of the redox peaks were sharper and higher than those of the cell with the polypropylene separator, indicating better battery kinetics of the cell with the high-temperature-resistant separator compared with the cell with the polypropylene separator. Similar results were also found for the cells prepared using the high-tem-perature-resistant separator without cellulose fibers.

The application of the high-temperature-resistant separator in the lithium-ion battery was also investigated. The pouch-type LiFePO$_4$/separator/graphite full cells with a high loading of LiFePO$_4$ (18.5 mg cm^{-2}) were assembled (Figure 5.58e and f). The initial discharge capacity of the cell with the high-temperature-resistant separa-tor was 135.8 mAh g^{-1} at 1 C, which was higher than that with the polypropylene separator (131.6 mAh g^{-1}) (Figure 5.58e). In addition, the capacity retention of the cell with the high-temperature-resistant separator was 92.3%, higher than that with the polypropylene separator (91.1%). The cell with the high-temperature-resistant separator showed smaller polarization compared with the cell with the polypropylene separator. Furthermore, the electrochemical impedance spectra of the full cells before and after recycling indicated that before recycling the cell with the high-temperature-resistant separator possessed a much lower bulk resistance (~5 Ω) compared with the polypropylene separator (14 Ω), implying faster ion diffusion. The cells exhibited two semicircles after recycling. The first semicircle at the high-frequency range was related with the formation of the insulative layer between the separator and electrodes, while the second semicircle corresponded to the charge-transfer resistance of the cell. The cell with the high-temperature-resistant separator possessed a much lower surface film resistance (~7.5 Ω) in comparison with that of the polypropylene separator (~17 Ω). This could be explained by the high affinity of the high-temperature-resistant separator to the liquid electrolyte, resulting in more uniformly wetted and stable interfaces in the cell, and thus inhibiting the growth of the undesirable insulative film. In addition, the cell with the high-temperature-resistant separator showed a better charge-transfer behavior than that with the polypropylene separator, which could promote the Faradaic reaction and alleviate the polarization of the cell. The recycling tests with increased number of cycles for the full batteries under a higher current density (2 C) could further confirm the enhanced recycling performance and rate capability of the high-temperature-resistant separator compared with the polypropylene separator, as shown in Figure 5.59.

The open-circuit voltage can be used to monitor the battery conditions at high temperatures. Once severe shrinkage happens to a separator, the cathode and anode will contact directly, leading to a sudden decrease in the open-circuit voltage. Figure 5.60a and b shows the open-circuit voltage curves of the LiFePO$_4$/separator/Li batteries with the high-temperature-resistant separator, and with the polypropylene separator at 150°C, and the battery device used for high-temperature measurement. The open-circuit voltage of the cell with the polypropylene separator rapidly decreased

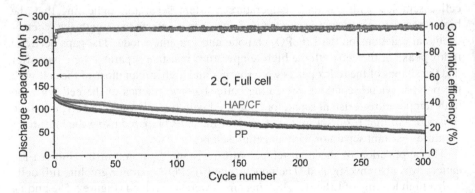

FIGURE 5.59 Recycling performance of the LiFePO$_4$/separator/graphite full battery with the high-temperature-resistant separator in comparison with the battery with the polypropylene separator at a current density of 2 C for 300 cycles. (Reprinted with permission from reference [85])

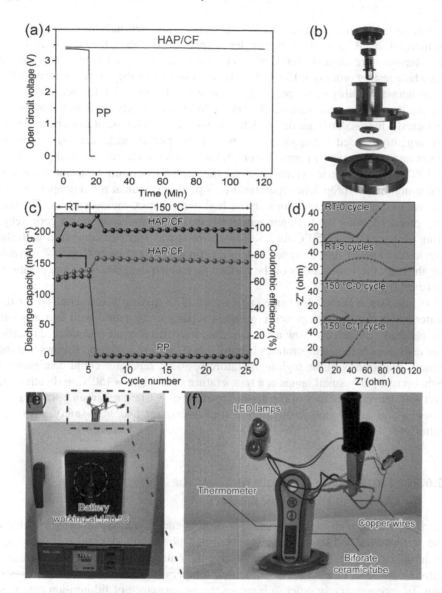

FIGURE 5.60 Electrochemical properties of the LiFePO₄/separator/Li batteries at 150°C. (a) Open-circuit voltage curves of the LiFePO₄/separator/Li batteries with the high-temperature-resistant separator, and with the polypropylene separator at 150°C; (b) digital image of the battery device used for high-temperature tests; (c) recycling performance of the batteries with the high-temperature-resistant separator, and with the polypropylene separator at 2 C; (d) electrochemical impedance spectra of the battery with the high-temperature-resistant separator; (e) the battery prepared using the high-temperature-resistant separator was continuously working at 150°C; (f) enlargement of the labeled region in (e). (Reprinted with permission from reference [85])

to zero in less than 20 minutes at 150°C because of the internal short circuit caused by thermal shrinkage of the polypropylene separator. However, the battery using the high-temperature-resistant separator could maintain its initial voltage throughout the whole testing process at 150°C for 2 hours. Because of the safety concern for the poor thermal stability of the polyolefin separators, the operation temperature of the conventional lithium-ion batteries is strictly limited normally to room temperature. In fact, the high-safety batteries which can work at high temperatures above 100°C are urgently needed for applications in various special fields such as aerospace crafts and nuclear power plants. Figure 5.60c shows the electrochemical properties at 150°C of the batteries prepared using the high-temperature-resistant separator, and using the polypropylene separator. Five cycles were run at room temperature to ensure normal battery conditions before high-temperature measurements. The battery prepared using the high-temperature-resistant separator showed a good recycling performance at 150°C for 25 cycles, which was superior compared with the previous reports.[85] The average Coulombic efficiency was higher than 98%, revealing the high thermal stability of the battery assembled using the high-temperature-resistant separator.

In addition, a higher discharge capacity of 157.8 mAh g^{-1} was achieved for the battery with the high-temperature-resistant separator compared with that at room temperature (138 mAh g^{-1}) owing to increased ion diffusion rate and reduced interfacial resistance at high temperatures (Figure 5.60d). Figure 5.60e and f shows that the battery prepared with the high-temperature-resistant separator could continuously light up two 3.0 V small lamps at a temperature of as high as 150°C, indicating the excellent thermal stability of the high-temperature-resistant separator and the great potential for application in high-temperature-resistant and high-safety advanced batteries.[85]

5.6.2. FIRE-RESISTANT ELECTRODE MATERIAL FOR ADVANCED LITHIUM-ION BATTERY

In recent years, with rapid development of new energy and clean energy vehicles, the performance and safety requirements of new power batteries and energy-storage batteries have become higher; for example, the batteries should withstand various extreme working conditions, such as high temperatures, and should be fire resistant. In recent years, in order to improve the performance of lithium-ion batteries, many studies focused on the development and improvement of electrode materials and electrolytes, while there were few studies on the design of electrodes and battery structures, especially for key battery materials which can withstand extreme conditions. A good structure design for the ion and electron diffusion pathways in the entire electrode is extremely important. By optimizing the electrode structure, the conductivity of the electrode and its electrolyte infiltration performance can be improved, and the transmission of electrons and ions within the entire electrode can be promoted, leading to enhanced energy density, power density, electrochemical properties, and safety of the battery. However, it is a big challenge to obtain a thick electrode with good electron/ion transportation characteristics and high active material loading amount. In addition, the structural design of the electrode also plays an important role in improving the safety of the battery.

The author's research group developed a free-standing ultrahigh-capacity fire-resistant nanocomposite thick (1.35 mm) electrode consisting of ultralong hydroxy-apatite nanowires, which exhibited excellent electrochemical properties, high loading amount of active material (108 mg cm^{-2}), high areal capacity (16.4 mAh cm^{-2}), high thermal stability, and excellent fire resistance. The use of ultralong hydroxyapatite nanowires in the electrode provided excellent thermal stability (structural integrity up to 1,000°C and electrochemical activity up to 750°C), fire resistance, and wide working temperature range.[86]

The ultrahigh-capacity fire-resistant electrode was prepared by a simple electrostatic-assisted self-assembly approach using ultralong hydroxyapatite nanowires, Ketjen black nanoparticles, carbon fibers, and LiFePO$_4$ powder as starting materials. After the self-assembly and suction filtration processes, the LiFePO$_4$ particles were uniformly embedded in the highly conductive and porous networks of ultralong hydroxyapatite nanowires/Ketjen black nanoparticles/carbon fibers without using any polymer binder, leading to a free-standing hierarchically nanostructured ultrahigh-capacity fire-resistant electrode. Moreover, the ultrahigh-capacity fire-resistant electrode possessed greatly enhanced capacity and rate capability owing to fast electron/ion transporting boosted redox reaction kinetics. Furthermore, the ultrahigh-capacity fire-resistant electrode in combination with the fire-resistant ultralong hydroxyapatite nanowire-based separator gave the as-assembled lithium-ion battery a wide working temperature range, from room temperature to 160°C; and such high-safety battery based on ultralong hydroxyapatite nanowires in both the electrode and separator would even work at higher temperatures if the electrolyte can also resist high temperatures.

Considering the areal capacity and manufacturing processibility, 18 mg cm^{-2} is usually adopted as the active material loading amount in commercial LiFePO$_4$ cathodes. Therefore, the ultrahigh-capacity fire-resistant electrode with an active material loading of 18 mg cm^{-2} was prepared and investigated. A top-view SEM image of the ultrahigh-capacity fire-resistant electrode is shown in Figure 5.61a. LiFePO$_4$ particles were uniformly embedded in the porous networks of ultralong hydroxyapatite nanowires/Ketjen black nanoparticles/carbon fibers without obvious particle aggregation. Furthermore, many interconnected pores were observed, which could facilitate the permeation of liquid electrolyte into the electrode. A high-magnification SEM image (Figure 5.61b) and the corresponding energy-dispersive X-ray spectroscopy elemental mapping results (Figure 5.61c) indicate that ultralong hydroxyapatite nanowires, carbon fibers, Ketjen black nanoparticles, and LiFePO$_4$ particles were self-assembled into a hybrid porous architecture. The hybrid porous architecture could effectively promote fast and continuous electron and ion transportation inside the electrode, leading to fast redox reaction kinetics and superior electrochemical properties.

The thickness of the as-prepared ultrahigh-capacity fire-resistant electrode was ~270 μm. Carbon fibers were uniformly embedded in the entire electrode, mimicking

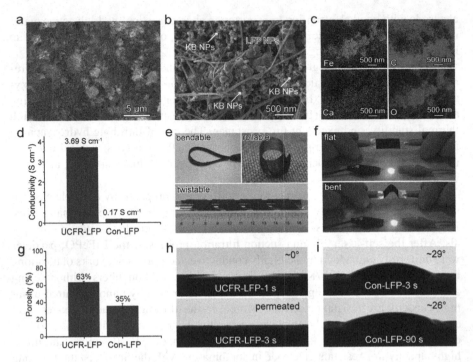

FIGURE 5.61 Characterization of the ultrahigh-capacity fire-resistant electrode (UCFR-LFP) and the conventional LiFePO$_4$ (Con-LFP) electrode. (a, b) SEM images (top view) of the ultrahigh-capacity fire-resistant electrode with different magnifications; (c) Fe, C, Ca, and O energy-dispersive X-ray spectroscopy elemental mapping images; (d) electronic conductivities; (e) digital images of the ultrahigh-capacity fire-resistant electrode with high flexibility; (f) digital images of an operating light emitting diode connected with the ultrahigh-capacity fire-resistant electrode at flat and bent states; (g) porosities; (h, i) contact angles of a liquid electrolyte on the ultrahigh-capacity fire-resistant electrode (h) and conventional LiFePO$_4$ electrode (i). (Reprinted with permission from reference [86])

a classical construction architecture with a steel-reinforced concrete structure. Similar to the reinforcing effect of the steel bars in the concrete, the embedded network-structured ultralong hydroxyapatite nanowires and carbon fibers could greatly enhance the mechanical strength of the electrode. In addition, carbon fibers with lengths of ~3 mm acted as high-speed electron transporting pathways embedded in all directions of the matrix that could further increase the electronic conductivity of the electrode. SEM images indicated that there were many layered gaps and interconnected pores in the ultrahigh-capacity fire-resistant electrode. The interconnected pores could facilitate the permeation of liquid electrolyte into all parts of the electrode through the capillary attraction, and the layered gaps could serve as electrolyte reservoirs inside the electrode and enable high electrolyte adsorption capacity, leading to enhanced electrochemical performance. In comparison, the conventional LiFePO$_4$ electrode with the same active material loading (18 mg cm^{-2}) was prepared by the typical slurry-coating method. After drying, LiFePO$_4$ particles and Super

P carbon additives were randomly bonded together by the polyvinylidene fluoride binder, forming an electrode film with a thickness of ~150 μm. Unlike the ultrahigh-capacity fire-resistant electrode with a large number of interconnected pores and layered gaps, only a few particles-stacked discrete voids were observed inside the conventional $LiFePO_4$ electrode. In addition, some particle aggregates were found on the surface of the conventional $LiFePO_4$ electrode, probably because of the poor compatibility between the organic-based polymer binder and the inorganic particles. It was also observed that $LiFePO_4$ particles were completely shielded by the polymer binder layer in some areas, and the dense and insulated polymer binder layers could block the electronic and ionic connections of $LiFePO_4$ particles, thus causing a negative effect to the electrochemical performance.[86]

Rapid electron transportation is a prerequisite for superior electrode performance. Four-point probe electronic conductivity measurements were taken to evaluate the electron transporting behavior of the ultrahigh-capacity fire-resistant electrode. As shown in Figure 5.61d, the electronic conductivity of the ultrahigh-capacity fire-resistant electrode was 3.69 S cm^{-1}, which was more than 20 times that of the conventional $LiFePO_4$ electrode (0.17 S cm^{-1}), and it was much higher than that of the fire-resistant electrode without carbon fibers (0.72 S cm^{-1}).

An ideal electrode should also be robust enough to withstand the stresses upon battery assembling and accidental collisions. The tensile strength of the ultrahigh-capacity fire-resistant electrode was 3.1 MPa, which was about three times that of the electrode without carbon fibers (1.1 MPa). The experimental results indicated the important role of the carbon fibers in enhancing both the electronic conductivity and the mechanical strength of the ultrahigh-capacity fire-resistant electrode. Figure 5.61e shows digital images of the ultrahigh-capacity fire-resistant electrode under different physical deformations, and it could be bent, rolled, and twisted without any visible damage, indicating good mechanical strength and high flexibility of the ultrahigh-capacity fire-resistant electrode. In addition, the brightness change could not be observed for the light emitting diode wired with the ultrahigh-capacity fire-resistant electrode under continuous bending and flattening cycles (Figure 5.61f), indicating the high flexibility and good electronic conductivity of the ultrahigh-capacity fire-resistant electrode even under bending conditions.

In order to improve the electrochemical reaction kinetics and fast electron transportation of the electrode, rapid and uniform ion diffusion inside all parts of the electrode is required. High porosity and well-defined pore structure are favorable for liquid electrolyte adsorption, permeability, and ion diffusion of the electrode. Figure 5.61g shows that the ultrahigh-capacity fire-resistant electrode possessed a much higher porosity (63%) than the conventional $LiFePO_4$ electrode (35%). The bubble-point pore size measurements indicated that the pores inside the ultrahigh-capacity fire-resistant electrode were well interconnected, with an average pore size of ~101.5 nm. The contact angle images of the ultrahigh-capacity fire-resistant electrode and conventional $LiFePO_4$ electrode with liquid electrolyte droplets are shown in Figure 5.61h and i. The conventional $LiFePO_4$ electrode showed a larger initial contact angle (29°) than that of the ultrahigh-capacity fire-resistant electrode (nearly 0°). Even after dropping the liquid electrolyte on the surface of the conventional

LiFePO$_4$ electrode for 90 seconds, the contact angle was still 26°. In contrast, the electrolyte droplet could completely permeate into the ultrahigh-capacity fire-resistant electrode after 3 seconds, showing the excellent electrolyte permeability of the ultrahigh-capacity fire-resistant electrode, which could promote the ionic transportation between the electrolyte and active material, thus promoting the electrochemical reaction kinetics.

The electrochemical properties of the ultrahigh-capacity fire-resistant electrode and the conventional LiFePO$_4$ electrode were investigated using coin-shaped cells with Li metal as both the counter and reference electrodes. The recycling properties and rate capabilities of the electrodes are shown in Figure 5.62a and b, respectively. At a current density of 0.5 C, the initial discharge capacity of the ultrahigh-capacity

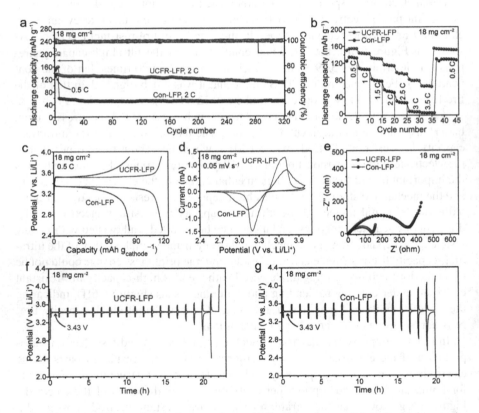

FIGURE 5.62 Electrochemical properties and kinetic analyses of the ultrahigh-capacity fire-resistant electrode (UCFR-LFP) and conventional LiFePO$_4$ electrode (Con-LFP). (a) Recycling performances at 2 C (6 mA cm^{-2}); the cells were stabilized at 0.5 C for 5 cycles before recycling at 2 C; (b) rate capabilities at various current densities ranging from 0.5 C to 3.5 C; (c) charge-discharge curves at 0.5 C obtained based on the capacities per unit weight of the entire electrode; (d) cyclic voltammetry curves at a sweep rate of 0.05 mV s^{-1}; (e) electrochemical impedance spectroscopy curves before recycling; (f, g) galvanostatic intermittent titration technique profiles of the ultrahigh-capacity fire-resistant electrode (f) and conventional LiFePO$_4$ electrode (g). (Reprinted with permission from reference [86])

fire-resistant electrode was 149.5 mAh g^{-1}, which was higher than that of the conventional LiFePO$_4$ electrode (130.4 mAh g^{-1}). With increasing current density from 0.5 C to 2 C, the capacity drop was not significant for the ultrahigh-capacity fire-resistant electrode, whereas a sharp decrease was observed for the conventional LiFePO$_4$ electrode (Figure 5.62a). After 315 cycles at 2 C, the ultrahigh-capacity fire-resistant electrode still showed a discharge capacity of 108 mAh g^{-1}, which was more than two times that of the conventional LiFePO$_4$ electrode (48.8 mAh g^{-1}), exhibiting the superior recycling performance of the ultrahigh-capacity fire-resistant electrode. When the current density increased from 0.5 C to 1 C, 1.5 C, 2.5 C, 3 C, and 3.5 C (10.5 mA cm^{-2}), the capacity of the ultrahigh-capacity fire-resistant electrode slowly decreased from 154.2 mAh g^{-1} to 143.7 mAh g^{-1}, 131.2 mAh g^{-1}, 115.8 mAh g^{-1}, 95 mAh g^{-1}, and 75.4 mAh g^{-1}, respectively, indicating its good rate capability. However, the rate capability of the conventional LiFePO$_4$ electrode was quite inferior, with a maximum capacity of 134.4 mAh g^{-1} at 0.5 C, and rapidly decreased to near zero when the current density increased above 3 C.[86]

The experiments indicated that the overpotentials of the ultrahigh-capacity fire-resistant electrode at various current densities were much smaller than those of the conventional LiFePO$_4$ electrode. The electrochemical performance tests showed that the electrode in the absence of LiFePO$_4$ particles showed negligible capacity. Benefiting from the removal of the metal foil current collector, the total areal mass of the ultrahigh-capacity fire-resistant electrode could be greatly reduced compared with that of the conventional LiFePO$_4$ electrode with similar loading amount of the active material. Therefore, an additional increase in the capacity per unit weight was anticipated for the ultrahigh-capacity fire-resistant electrode. As shown in Figure 5.62c, the ultrahigh-capacity fire-resistant electrode showed a much higher discharge capacity (118.5 mAh g^{-1}) than the conventional LiFePO$_4$ electrode (82.4 mAh g^{-1}) at 0.5 C obtained based on the capacity per unit weight of the entire cathode.

The cyclic voltammetry curves of the ultrahigh-capacity fire-resistant electrode and conventional LiFePO$_4$ electrode are shown in Figure 5.62d. There were two typical peaks in the cathodic and anodic scans for both samples due to lithium extraction and insertion processes of LiFePO$_4$, respectively. The gap between the redox peaks of the ultrahigh-capacity fire-resistant electrode was narrower than that of the conventional LiFePO$_4$ electrode, implying the superior electrochemical kinetics of the ultrahigh-capacity fire-resistant electrode. In addition, the electrochemical impedance spectroscopy and galvanostatic intermittent titration technique were used to evaluate the fast electron and ion transporting behaviors of the ultrahigh-capacity fire-resistant electrode. The Nyquist plots of the ultrahigh-capacity fire-resistant electrode and conventional LiFePO$_4$ electrode indicated that each electrode exhibited a single depressed semicircle in the high-medium frequency region (Figure 5.62e) due to the charge-transfer resistance. A much smaller charge-transfer resistance of the ultrahigh-capacity fire-resistant electrode was observed compared with that of the conventional LiFePO$_4$ electrode (140 Ω versus 350 Ω), indicating rapid charge transfer and ion transportation of the ultrahigh-capacity fire-resistant electrode. The galvanostatic intermittent titration technique profiles showed the quasi-equilibrium voltage evolution of the ultrahigh-capacity fire-resistant electrode and conventional

LiFePO$_4$ electrode during the repeated charging (discharging) and resting steps (Figure 5.62f and g). The static voltage of all resting steps in the charging and discharging processes was 3.43 V for both electrodes, indicating that the lithium insertion/extraction processes occurred at this potential. However, the ohmic polarization of the ultrahigh-capacity fire-resistant electrode in each charging and discharging step was smaller than that of the conventional LiFePO$_4$ electrode, indicating the lower electrode resistance of the ultrahigh-capacity fire-resistant electrode. The ionic diffusion coefficient was 3.6×10^{-11} cm^2 s^{-1} for the ultrahigh-capacity fire-resistant electrode and 0.5×10^{-11} cm^2 s^{-1} for the conventional LiFePO$_4$ electrode, further supporting the faster ionic transporting behavior of the ultrahigh-capacity fire-resistant electrode compared with the conventional LiFePO$_4$ electrode.

Thick battery electrodes with high loading amounts of active materials and areal capacities are important for enhanced energy density and lower manufacturing cost of the lithium-ion batteries. The experiments showed that the conventional slurry-coated LiFePO$_4$ electrode could not obtain an active material loading of 36 mg cm^{-2} due to many cracks on the surface of the conventional LiFePO$_4$ electrode because the binding strength of the polyvinylidene fluoride binder could not withstand high thermal shrinkage of the electrode in the drying process, which was also a common problem for conventional thick electrodes with high active material loadings.

As shown in Figure 5.63a–c, thick ultrahigh-capacity fire-resistant electrodes with different active material loadings (36 mg cm^{-2}, 72 mg cm^{-2}, and 108 mg cm^{-2}) and thicknesses (~460 μm, 900 μm, and 1350 μm, respectively) were prepared with no cracks or defects. Interconnected pores were observed in the ultrahigh-capacity fire-resistant electrode with an active material loading of 108 mg cm^{-2} (Figure 5.63d), implying fast electron and ion transportation even in the electrode with such a large thickness (1.35 mm). In addition, the ultrahigh-capacity fire-resistant electrode with an active material loading of 108 mg cm^{-2} possessed good mechanical properties with a tensile strength higher than 2 MPa and compressive strength up to 9 MPa, and the ultrahigh-capacity fire-resistant electrode with an active material loading of 108 mg cm^{-2} could preserve its integrity even with a big strain of 35% after the compression test with a large strain tolerance because of its unique pore structure (i.e., the interconnected pores and layered gaps) acting as buffer spaces for external stress, which could improve the battery safety in the case of accidental collisions.

Because of the high loading of active material, a great enhancement in areal capacity was obtained in the thick ultrahigh-capacity fire-resistant electrode. As shown in Figure 5.63e, the areal capacity increased proportionally with the active material loading with a maximum value of 15.9 mAh cm^{-2} at an active material loading of 108 mg cm^{-2} and a current density of 3.6 mA cm^{-2}, which was very high for the cathode materials of lithium-ion batteries. The corresponding areal and volumetric energy densities of the ultrahigh-capacity fire-resistant electrode with an active material loading of 108 mg cm^{-2} were 50.4 mWh cm^{-2} and 373 Wh L^{-1}, respectively, which were comparable to the previously reported results for LiFePO$_4$ cathode materials. The electrode conditions of the ultrahigh-capacity fire-resistant cathode with an active material loading of 108 mg cm^{-2} and lithium metal anode after recycling tests were further investigated. The ultrahigh-capacity fire-resistant electrode with

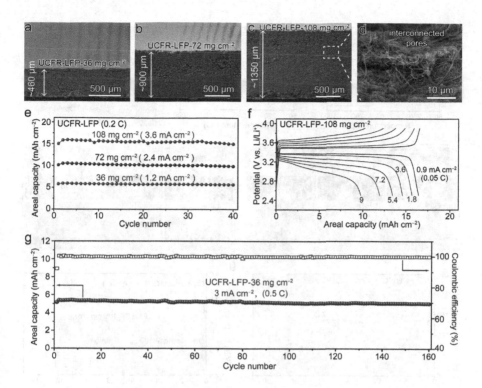

FIGURE 5.63 Morphologies and electrochemical performances of the ultrahigh-capacity fire-resistant electrodes with high loading amounts of the active material. (a–c) Cross-sectional SEM images of the ultrahigh-capacity fire-resistant electrodes with different active material loadings of 36 mg cm^{-2} (a), 72 mg cm^{-2} (b), and 108 mg cm^{-2} (c); (d) enlarged SEM image of (c); (e) recycling performances of the ultrahigh-capacity fire-resistant electrodes with different active material loadings; (f) charge-discharge curves of the ultrahigh-capacity fire-resistant electrode at various current densities ranging from 0.9 mA cm^{-2} to 9 mA cm^{-2}; (g) recycling performance of the ultrahigh-capacity fire-resistant electrode at a current density of 3 mA cm^{-2} (0.5 C). (Reprinted with permission from reference [86])

an active material loading of 108 mg cm^{-2} could be well maintained with good strain tolerance during the recycling tests with high-capacity and high-current-density delithiation and lithiation. Even at a current density of 9 mA cm^{-2}, the ultrahigh-capacity fire-resistant electrode with an active material loading of 108 mg cm^{-2} still exhibited a high areal capacity of ~10 mAh cm^{-2} (Figure 5.63f). As shown in Figure 5.63g, the ultrahigh-capacity fire-resistant electrode with an active material loading of 36 mg cm^{-2} showed a good recycling stability with an areal capacity of 5 mAh cm^{-2}, and the retention rate was up to 97% after 160 cycles at 3 mA cm^{-2}.

High safety is a prerequisite for batteries, especially for the ones with large sizes and high energy densities. Heat can be generated in a short time once the battery energy is released, which may result in fire and even explosion, and cause great threats to human life. Battery materials with high resistance to both fire and high

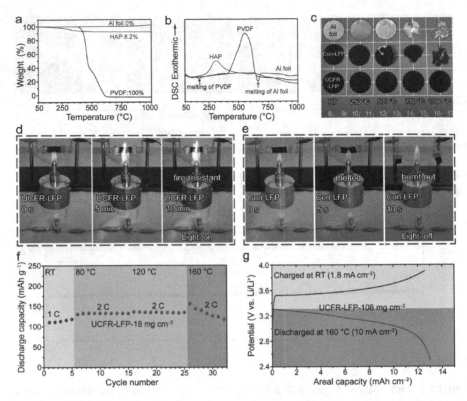

FIGURE 5.64 Thermal stability tests of the ultrahigh-capacity fire-resistant electrode and conventional LiFePO$_4$ electrode. (a) Thermogravimetric curves and (b) differential scanning calorimetry curves of ultralong hydroxyapatite nanowires, polyvinylidene fluoride (PVDF) binder, and aluminum foil current collector; (c) thermal stability tests of the aluminum foil, conventional LiFePO$_4$ electrode with an active material loading of 18 mg cm^{-2}, and ultrahigh-capacity fire-resistant electrode with an active material loading of 18 mg cm^{-2} at different temperatures; (d, e) fire-resistance tests of the ultrahigh-capacity fire-resistant electrode with an active material loading of 18 mg cm^{-2} (d) and the conventional LiFePO$_4$ electrode with an active material loading of 18 mg cm^{-2} (e); (f) recycling performance of the ultrahigh-capacity fire-resistant electrode with an active material loading of 18 mg cm^{-2} at temperatures ranging from room temperature to 160°C; (g) charge-discharge curves of the ultrahigh-capacity fire-resistant electrode with an active material loading of 108 mg cm^{-2}, the cell was charged at room temperature and discharged at 160°C. (Reprinted with permission from reference [86])

temperatures are very important to enhancing the safety and widening the working temperature range of batteries.

As shown in Figure 5.64a and b, a small and broad endothermic peak around 160°C was attributed to its softening and melting of the polyvinylidene fluoride binder. Further increasing the temperature to above 400°C, the polyvinylidene fluoride binder started to decompose, and was completely burned out at round 600°C. Although no weight loss was observed for the aluminum foil current collector in the thermogravimetric curve, the sharp and evident endothermic peak in the differential scanning calorimetry curve showed its melting at around 650°C, which could affect the workability of the electrode, and cause collateral damage to other parts of the battery. In contrast, no obvious peak below 250°C was observed for ultralong hydroxyapatite nanowires in the differential scanning calorimetry curve. Ultralong hydroxyapatite nanowires could maintain 91.8% of its initial weight even at a high temperature of 1,000°C, and the small weight loss (8.2%) was caused by the loss of the adsorbed water and oleic acid molecules. The thermogravimetric and differential scanning calorimetry analyses of the ultrahigh-capacity fire-resistant electrode indicated its excellent thermal stability.

As shown in Figure 5.64c, the ultrahigh-capacity fire-resistant electrode could well preserve its size stability and structural integrity without any visible damage after thermal treatment at a high temperature of 1,000°C in argon atmosphere. In contrast, the conventional $LiFePO_4$ electrode ruptured after thermal treatment at 500°C, and was completely ruined after heating at 1,000°C because of poor thermal stability of the polyvinylidene fluoride binder and aluminum foil current collector. Furthermore, no obvious change was observed for the crystalline phases in the ultrahigh-capacity fire-resistant electrode after thermal treatment up to 750°C. In addition, the ultrahigh-capacity fire-resistant electrode exhibited excellent fire-resistant performance (Figure 5.64d). In contrast, the conventional $LiFePO_4$ electrode was rapidly burned in fire and could not work after 10 seconds (Figure 5.64e). The fire-resistant performance of the ultrahigh-capacity fire-resistant electrode in the presence of a liquid electrolyte was also studied. Owing to the high adsorption capacity of the ultrahigh-capacity fire-resistant electrode for the liquid electrolyte, the liquid electrolyte could be rapidly adsorbed into the inner pores of the electrode, and the ignited liquid electrolyte was rapidly extinguished within seconds. The nonflammability of the ultrahigh-capacity fire-resistant electrode could greatly enhance the safety of the batteries. However, due to poor thermal stability of the conventional battery components, the working temperatures of lithium-ion batteries have long been restricted from room temperature to about 60°C. The high thermal stability of the ultrahigh-capacity fire-resistant electrode has a great potential in widening the operation temperature range of batteries. Combined with the thermally stable ultralong hydroxyapatite nanowires-based separator, the ultrahigh-capacity fire-resistant electrode with an active material loading of 18 mg cm^{-2} showed a wide operation (charging and discharging) temperature window ranging from room temperature to 160°C (Figure 5.64f). Even at a temperature of 160°C, the thick ultrahigh-capacity fire-resistant electrode with an active material loading of 108 mg cm^{-2} could still possess a high areal capacity of 13 mAh cm^{-2} at a current density of 10 mA cm^{-2} (Figure 5.64g). The stable operation at 160°C of the battery assembled with the ultrahigh-capacity fire-resistant electrode with an active material loading of 108 mg cm^{-2} and ultralong hydroxyapatite nanowires-based separator was also realized.[86]

5.6.3. HIGH-TEMPERATURE-RESISTANT SEPARATOR AND ELECTRODE MATERIAL FOR HIGH-ENERGY-DENSITY LITHIUM-ION CAPACITOR

The ongoing surge in demand for sustainable energy supplies and ever-increasing environmental concerns have driven the development of clean and efficient energy-storage devices. With a fast-growing market for consumer electronics, electric vehicles, and smart grids, there is also a strong demand for energy-storage devices with high energy density, high power density, long recycling life, and high safety. Among various electrochemical energy-storage systems, lithium-ion batteries and supercapacitors are two typical devices that have already been commercialized and widely used nowadays. Because of their different charge-storage mechanisms, lithium-ion batteries and supercapacitors show complementary energy-storage functions. For example, lithium-ion batteries are able to deliver high working voltage and high energy density (150 Wh kg^{-1}–200 Wh kg^{-1}), but usually suffer from low power density (below 1 kW kg^{-1}) and poor recycling performance (< 1,000 cycles). However, supercapacitors usually possess a high power density (> 10 kW kg^{-1}), long recycling lifetime (> 10,000 cycles), and fast charge/discharge processes (within seconds), while limited by the low energy density (< 10 Wh kg^{-1}). Lithium-ion capacitors (also called as lithium-ion hybrid supercapacitors) have promising applications with the advantage of combining the complementary features of lithium-ion batteries and supercapacitors with both high power density and high energy density. Usually, lithium-ion capacitors can be constructed from a battery-type anode and a supercapacitor-type cathode with a lithium-salt-containing electrolyte. With increased charging/discharge rates, the materials with high thermal stability are crucial for high-safety lithium-ion capacitors.

A high-temperature-resistant high power density high-energy-density lithium-ion capacitor made from ultralong hydroxyapatite nanowires-based flexible electrodes and separator was developed using the electron/ion dual highly conductive and fire-resistant composite Li$_4$Ti$_5$O$_{12}$-based anode and activated carbon-based cathode, together with a fire-resistant ultralong hydroxyapatite nanowire separator.[87] The aqueous slurry of the electrode was prepared by mixing active materials (Li$_4$Ti$_5$O$_{12}$ or activated carbon), Ketjen black nanoparticles, carbon fibers, and ultralong hydroxyapatite nanowires, and a self-supported flexible cathode or anode membrane was obtained after the filtration on the filter paper and drying process. The digital image of the as-prepared flexile high-temperature-resistant electrode with a diameter of 4 cm is shown in Figure 5.65. The bending test showed that the as-prepared high-temperature-resistant electrode possessed a good flexibility and mechanical strength (~2.5 MPa), which is promising for the application in the portable and wearable devices. In the high-temperature-resistant lithium-ion capacitor, the fire-resistant ultralong hydroxyapatite nanowires were the important component of all parts including the cathode, anode, and separator. Such unique cathode and anode consisting of ultralong hydroxyapatite nanowires possessed continuous and fast charge-delivery behaviors, high thermal stability and durability, and excellent fire resistance. The as-prepared high-temperature-resistant lithium-ion capacitor showed superior electrochemical properties especially at high current density and high temperatures, with a long recycling lifetime, high areal energy density (1.58 mWh cm^{-2}) at a high active material loading of 30 mg cm^{-2}, and a wide working temperature range from room temperature to 150°C.

The porous structure and interconnected pores of the flexile high-temperature-resistant electrode could provide continuous channels for liquid electrolyte diffusion,

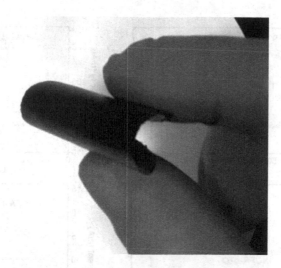

FIGURE 5.65 A digital image of the as-prepared flexile high-temperature-resistant electrode consisting of ultralong hydroxyapatite nanowires with a thickness of ~120 μm. (Reprinted with permission from reference [87])

and the pores and interspaces between layers could act as reservoirs for adsorption of liquid electrolytes. The experiments indicated that the liquid electrolyte drop could rapidly spread out and penetrate into the flexile high-temperature-resistant electrode membrane within seconds, exhibiting excellent wettability to the liquid electrolyte and fast ion diffusion property. In comparison, the conventional $Li_4Ti_5O_{12}$ electrode prepared by the slurry-coating method possessed no obvious porous structure and exhibited poor wettability to the liquid electrolyte. SEM observations showed some aggregations and cracks in the conventional $Li_4Ti_5O_{12}$ electrode, and some $Li_4Ti_5O_{12}$ particles were encapsulated by the polyvinylidene fluoride binder, which could weaken the electrochemical performance of the electrode.

The electrochemical properties of the as-prepared flexile high-temperature-resistant $Li_4Ti_5O_{12}$-based anode consisting of ultralong hydroxyapatite nanowires and conventional $Li_4Ti_5O_{12}$ anode with the same active material loading are shown in Figure 5.66a–d. The half cells were assembled using the ultralong hydroxyapatite nanowire separator and the Li foil as both the counter and reference electrode. The high-temperature-resistant $Li_4Ti_5O_{12}$-based anode showed an average discharge capacity of ~96 mAh g^{-1} at 4 A g^{-1} for 300 cycles (Figure 5.66a), which was much higher than that of the conventional $Li_4Ti_5O_{12}$ anode (47 mAh g^{-1}). Figure 5.66b shows the rate capabilities of the high-temperature-resistant $Li_4Ti_5O_{12}$-based anode and conventional $Li_4Ti_5O_{12}$ anode. The average discharge capacities of the high-temperature-resistant $Li_4Ti_5O_{12}$-based anode at the current densities of 0.5 A g^{-1}, 1 A g^{-1}, 2 A g^{-1}, 3 A g^{-1}, and 4 A g^{-1} were 209 mAh g^{-1}, 152 mAh g^{-1}, 116 mAh g^{-1}, 101 mAh g^{-1}, and 88 mAh g^{-1}, respectively, which were much higher than those of the conventional $Li_4Ti_5O_{12}$ anode at the same current densities. Even at a high current density of 6 A g^{-1}, the high-temperature-resistant $Li_4Ti_5O_{12}$-based anode still possessed a

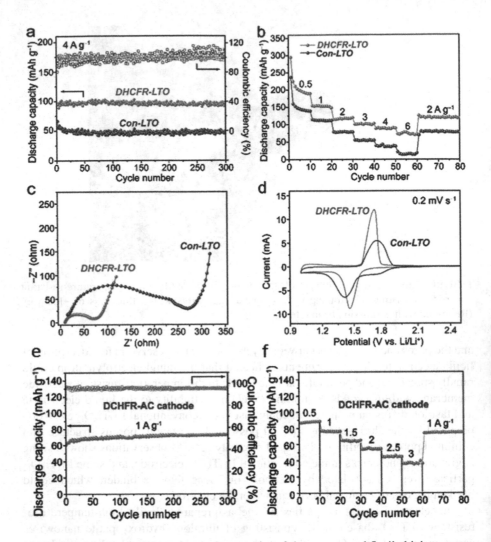

FIGURE 5.66 (a–d) Electrochemical properties of the as-prepared flexile high-tempera-ture-resistant $Li_4Ti_5O_{12}$-based anode consisting of ultralong hydroxyapatite nanowires, and conventional $Li_4Ti_5O_{12}$ anode: (a) recycling performance at a current density of 4 A g^{-1}; (b) rate capabilities under different current densities ranging from 0.5 A g^{-1} to 6 A g^{-1}; (c) electrochemical impedance spectra before recycling; (d) cyclic voltammogram profiles at a sweep rate of 0.2 mV s^{-1}; the cells were assembled using the ultralong hydroxyapatite nanowire fire-resistant separator with Li metal as both the counter and reference electrode; (e, f) recycling performance and rate capability of the as-prepared flexible high-temperature-resistant activated carbon-based cathode consisting of ultralong hydroxyapatite nanowires; the cells were assembled with the ultralong hydroxyapatite nanowire fire-resistant separa-tor, and tested at the potential window of 2.5 V–4.2 V versus Li/Li+. (Reprinted with per-mission from reference [87])

high discharge capacity of 73 mAh g^{-1}, however, the discharge capacity of the conventional $Li_4Ti_5O_{12}$ anode rapidly decreased to lower than 10 mAh g^{-1} at 6 A g^{-1} (Figure 5.66b). As shown in the electrochemical impedance spectra (Figure 5.66c), the charge-transfer resistance (semicircle in the high-medium frequency range) of the high-temperature-resistant $Li_4Ti_5O_{12}$-based anode (~70 Ω) was much lower than that of the conventional $Li_4Ti_5O_{12}$ anode (~270 Ω). The cyclic voltammogram results indicated that the shape of the redox peaks of the high-temperature-resistant $Li_4Ti_5O_{12}$-based anode was much sharper (Figure 5.66d), and the polarization-potential gap between the redox peaks of the high-temperature-resistant $Li_4Ti_5O_{12}$-based anode was smaller than that of the conventional $Li_4Ti_5O_{12}$ anode. These experimental results showed enhanced electrochemical kinetics of the high-temperature-resistant $Li_4Ti_5O_{12}$-based anode compared with the conventional $Li_4Ti_5O_{12}$ anode.

Figure 5.66e and f shows the electrochemical properties of the as-prepared high-temperature-resistant activated carbon-based cathode. The high-temperature-resistant activated carbon-based cathode also showed high electronic conductivity and ion accessibility, and a discharge capacity of ~70 mAh g^{-1} at a current density of 1 A g^{-1} for 300 cycles (Figure 5.66e). The discharge capacities of the high-temperature-resistant activated carbon-based cathode were ~86.7 mAh g^{-1}, 76.8 mAh g^{-1}, 64.9 mAh g^{-1}, 54 mAh g^{-1}, and 44.7 mAh g^{-1} at a current density of 0.5 A g^{-1}, 1 A g^{-1}, 1.5 A g^{-1}, 2 A g^{-1}, and 2.5 A g^{-1}, respectively. Even at a high current density of 3 A g^{-1}, the high-temperature-resistant activated carbon-based cathode could still possess a high discharge capacity of 36.1 mAh g^{-1} with a capacity retention of 42%, exhibiting the excellent rate capability.[87]

Figure 5.67 shows electrochemical properties of the ultralong hydroxyapatite nanowires-based high-temperature-resistant lithium-ion capacitor. Figure 5.67a shows the structure of the ultralong hydroxyapatite nanowires-based high-temperature-resistant lithium-ion capacitor, which was assembled with a high-temperature-resistant $Li_4Ti_5O_{12}$-based anode, a high-temperature-resistant separator, and a high-temperature-resistant activated carbon-based cathode, all three parts (anode, cathode, separator) contained ultralong hydroxyapatite nanowires. The high-temperature-resistant lithium-ion capacitor possessed superior electron/ion transporting behaviors and electrochemical properties. The galvanostatic charge and discharge curves of the high-temperature-resistant lithium-ion capacitor are shown in Figure 5.67b, which exhibited a typical triangle-shaped profile. The specific cell capacitances of the high-temperature-resistant lithium-ion capacitor were ~85.8 F g^{-1}, 81.4 F g^{-1}, 74.7 F g^{-1}, 65.5 F g^{-1}, and 53 F g^{-1} at a current density of 0.1 A g^{-1}, 0.2 A g^{-1}, 0.5 A g^{-1}, 1 A g^{-1}, and 2 A g^{-1}, respectively.

One of the important properties of the lithium-ion capacitor is its cycling lifetime. Unfortunately, the cycling lifetimes of many previous reported lithium-ion capacitor using the $Li_4Ti_5O_{12}$-based anodes and activated carbon-based cathodes were only thousands of cycles. For the high-temperature-resistant lithium-ion capacitor, a high capacity retention of 94.5% was achieved even upon recycling over 34,000 cycles (Figure 5.67c). The Ragone plot (Figure 5.67d) indicated that a maximum energy density of 71.54 Wh kg^{-1} could be obtained for the high-temperature-resistant lithium-ion capacitor at a power density of 150 W kg^{-1}. Even at a high power density of 7.5 kW kg^{-1} (with charging or discharging time for 9 seconds), the

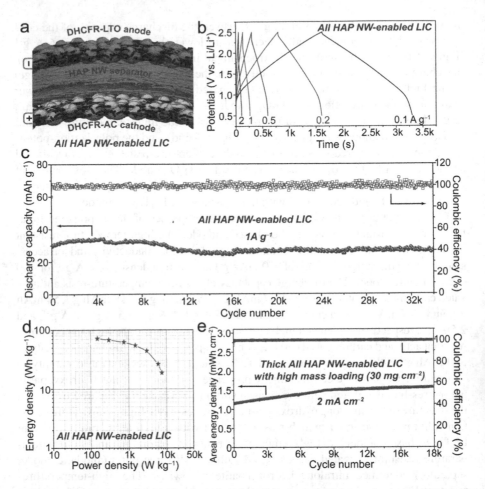

FIGURE 5.67 Electrochemical properties of the ultralong hydroxyapatite nanowires-based high-temperature-resistant lithium-ion capacitor. (a) A schematic illustration for the integrated design concept of a high-temperature-resistant lithium-ion capacitor assembled with the high-temperature-resistant $Li_4Ti_5O_{12}$-based anode, a high-temperature-resistant separator, and a high-temperature-resistant activated carbon-based cathode; (b) galvanostatic charge-discharge curves at different current densities ranging from 0.1 A g^{-1} to 2 A g^{-1}; (c) recycling performance; (d) Ragone plot; (e) recycling performance and areal energy density of the high-temperature-resistant lithium-ion capacitor made of the high-temperature-resistant $Li_4Ti_5O_{12}$-based anode and high-temperature-resistant activated carbon-based cathode with a commercial-level loading of the active materials (30 mg cm^{-2}). (Reprinted with permission from reference [87])

high-temperature-resistant lithium-ion capacitor still possessed an energy density of 18.75 Wh kg^{-1}. The electrochemical properties of the high-temperature-resistant lithium-ion capacitor (e.g., energy density 71.54 Wh kg^{-1}, power density 7.5 kW kg^{-1}, and capacity retention 94.5% over 34,000 cycles) were superior to many previous reports on lithium-ion capacitors with the same class of anodes and cathodes. Figure 5.67e shows the electrochemical performance of the high-temperature-resistant

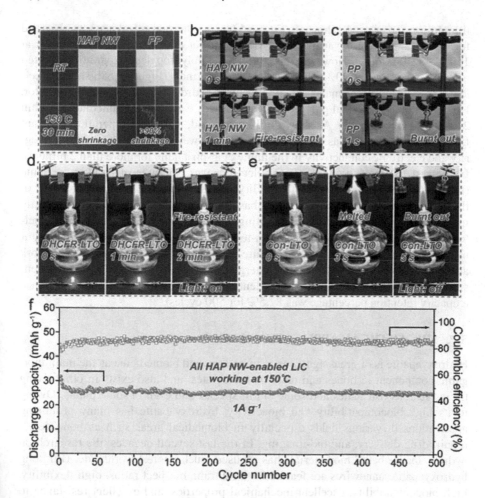

FIGURE 5.68 Thermal stability, fire resistance, and high-temperature recycling performance of the as-prepared high-temperature-resistant lithium-ion capacitor. (a) Thermal stability of the ultralong hydroxyapatite nanowire separator and the commercial polypropylene separator; (b, c) fire-resistant tests of the ultralong hydroxyapatite nanowire separator and the commercial polypropylene separator; (d, e) fire-resistant tests of the high-temperature-resistant $Li_4Ti_5O_{12}$-based anode and the conventional $Li_4Ti_5O_{12}$ electrode, the electrode was fixed above the flame of a spirit lamp and connected with a working light emitting diode; (f) high-temperature recycling performance of the high-temperature-resistant lithium-ion capacitor at 150°C. (Reprinted with permission from reference [87])

lithium-ion capacitor with a high loading of the active materials in the electrodes (total 30 mg cm^{-2}: 10 mg cm^{-2} for the anode and 20 mg cm^{-2} for the cathode). A very high areal energy density of 1.58 mWh cm^{-2} was obtained for the high-temperature-resistant lithium-ion capacitor.

Figure 5.68 shows the excellent thermal stability, fire resistance and high-temperature recycling performance of the as-prepared high-temperature-resistant lithium-ion capacitor. As shown in Figure 5.68a, no thermal shrinkage was observed for the ultralong hydroxyapatite nanowire separator after the thermal treatment at 150°C for 30 minutes. However, the commercial polypropylene separator exhibited a severe shrinkage (> 90%) after the thermal treatment. Figure 5.68b shows excellent fire-resistant performance of the ultralong hydroxyapatite nanowire separator. However, the commercial polypropylene separator was burned immediately within 2 seconds (Figure 5.68c). As shown in Figure 5.68d, the high-temperature-resistant $Li_4Ti_5O_{12}$-based anode was fire-resistant and could well maintain its structural integrity and electronic conductivity in the flame of a spirit lamp for more than 2 minutes. In contrast, the conventional $Li_4Ti_5O_{12}$ electrode was rapidly burned out in the flame (Figure 5.68e). The high thermal stability and excellent fire-resistant property of the ultralong hydroxyapatite nanowires-based separator and electrodes could greatly improve the safety and extend the operating temperature range of lithium-ion capacitors. As shown in Figure 5.68f, the high-temperature-resistant lithium-ion capacitor using a thermally stable room-temperature ionic liquid-based electrolyte could work at a temperature of 150°C with a high energy density of 45.5 Wh kg^{-1} at a power density of 1,500 W kg^{-1} and exhibited excellent high-temperature recycling performance (capacity retention percentage was 82.5% for 500 cycles).[87]

5.7. BIOMEDICAL APPLICATIONS

Hydroxyapatite is of great significance in biology and biomedicine as the main inorganic component in bones and teeth of vertebrates, and also exists in other living things such as fish scales and some chitin species and even in some plants. Owing to its high biocompatibility and bioactivity, hydroxyapatite has many promising applications in various fields especially in biomedical areas such as bone defect repair, drug delivery, and bio-imaging. In the last several decades, the research on hydroxyapatite-based biomaterials has aroused much interest worldwide. Ultralong hydroxyapatite nanowires are featured with ultrahigh aspect ratios, high flexibility, high biocompatibility, excellent mechanical properties, and excellent resistance to fire and high temperatures. Thus, ultralong hydroxyapatite nanowires are promising for applications in various biomedical fields.

Deformable biomaterials with excellent flexibility, softness, or elasticity are very important for biomedical applications. Overviewing deformable biomaterials reported in the literature, they were mainly polymers and polymer-based composites. Deformable biomaterials based on inorganic materials such as hydroxyapatite were rarely reported due to their high brittleness.

5.7.1. DEFORMABLE BIOMATERIALS BASED ON ULTRALONG HYDROXYAPATITE NANOWIRES

It is well known that hydroxyapatite biomaterials are usually hard and brittle, which may restrict their applications. How to overcome the high brittleness of hydroxyapatite biomaterials? Hybridizing hydroxyapatite with soft polymers is a useful strategy to prepare deformable composite materials, in which the deformable function is mainly contributed by organic polymers. As discussed above, the author's research group found that ultralong hydroxyapatite nanowires with diameters of < 100 nm and lengths of > 100 μm are highly flexible, and thus can overcome high brittleness of hydroxyapatite materials.[12] In addition, the calcium oleate precursor solvothermal/hydrothermal method was developed for the synthesis of ultralong hydroxyapatite nanowires, and ultralong hydroxyapatite nanowires with diameters of ~10 nm and lengths up to hundreds of micrometers with ultrahigh aspect ratios up to > 10,000 and high flexibility were successfully prepared by this method. By optimizing the synthetic conditions, for example, by using methanol instead of ethanol in the calcium oleate precursor solvothermal method, the lengths of the as-prepared ultralong hydroxyapatite nanowires could reach nearly one millimeter.[31]

This author proposed a new concept of deformable biomaterials consisting of ultralong hydroxyapatite nanowires based on their high flexibility, which have unique advantages such as high flexibility, softness, or elasticity.[88] Deformable biomaterials based on ultralong hydroxyapatite nanowires can overcome the high brittleness and high hardness of the traditional hydroxyapatite biomaterials. The deformable properties, which are usually found in polymer-based materials, will greatly extend applications of ultralong hydroxyapatite nanowires-based biomaterials.[88] For a detailed review and discussion on deformable biomaterials based on ultralong hydroxyapatite nanowires, readers can refer to a recent review article written by the author.[88] This review article presented a brief review on the recent progress on deformable biomaterials based on ultralong hydroxyapatite nanowires and proposed a strategy for solving the high brittleness of the traditional hydroxyapatite biomaterials using ultralong hydroxyapatite nanowires, and a variety of deformable biomaterials based on ultralong hydroxyapatite nanowires were discussed, including highly flexible biomedical paper, highly flexible antibacterial biomedical paper, highly flexible photoluminescent paper, highly ordered ultralong hydroxyapatite nanowires and their derived highly flexible biomaterials, elastic porous biomaterials, elastic hydrogels, highly porous elastic aerogel, and other deformable biomaterials. Ultralong hydroxyapatite nanowires-based deformable biomaterials have many advantages such as excellent deformability, high biocompatibility/bioactivity, good biodegradation, excellent ability for cell adhesion/spreading/proliferation, excellent osteoinduction/osteogenesis/osseointegration/neovascularization, excellent ability of loading and releasing drugs/proteins/growth factors, etc. Deformable biomaterials based on ultralong hydroxyapatite nanowires will extend the range of biomedical applications of hydroxyapatite biomaterials to various fields such as bone regeneration, artificial periosteum, skin wound healing, biomedical paper, medical test paper, drug delivery, diagnosis, and therapy.[88]

5.7.2. HIGH-PERFORMANCE BIOMEDICAL PAPER

As discussed above, hydroxyapatite is the main inorganic component of vertebrate bones and teeth, which has excellent biocompatibility and biological activity and good application prospects in the field of biomedicine. However, biomaterials composed of a single phase of hydroxyapatite are generally brittle and have no flexibility, and are difficult to mold into specific shapes required for various biomedical applications. In addition, deformable biomaterials are required in some specific biomedical applications. For this reason, the design and synthesis of nanocomposite biomaterials comprising ultralong hydroxyapatite nanowires and biopolymers with good flexibility and excellent mechanical properties are promising for practical applications. Compared with the metallic implants, the hydroxyapatite/biocompatible polymer composite materials exhibit good biodegradation and adjustable mechanical properties, and can avoid post-implantation surgeries. Hydroxyapatite nanowire-reinforced nanocomposite biomaterials are highly desirable for bone tissue engineering with superior reinforcement performance and biological responses. As the length of the hydroxyapatite nanowires increases, the mechanical properties of the nanocomposite materials will be enhanced owing to increasing interactions, including overlapping, winding, crossing, strengthening, and hydrogen bonding.

(1) High-Performance Biomedical Paper Comprising Ultralong Hydroxyapatite Nanowires and Chitosan

Chitosan is an amino-polysaccharide, a deacetylated derivative of chitin, which is a natural component of the shells of crustaceans, and is the second most abundant natural polymer with high biocompatibility, good degradability, and antibacterial properties. Chitosan is a linear polymer composed of glucosamine (2-amino-2-deoxy-d-glucopyranose), and is a cationic polymer that is soluble in weakly acidic solution. Chitosan is a promising biopolymer for applications in drug delivery, theranostics, gene therapy, and tissue engineering, and can be used as a hemostatic agent and skin wound dressing to promote blood coagulation and wound healing. Therefore, it is expected that hybridizing ultralong hydroxyapatite nanowires with a biopolymer such as chitosan or collagen will obtain high-performance deformable biomaterials with excellent properties.

The author's research group developed a highly flexible multifunctional biomedical paper comprising ultralong hydroxyapatite nanowires and chitosan with a wide weight percentage range of ultralong hydroxyapatite nanowires up to 100 wt.%.[89] The as-prepared biomedical paper exhibited a high flexibility and superior mechanical properties. The surface wettability, swelling ratio, and water vapor transmission rate of the biomedical paper could be controlled by adjusting the addition amount of ultralong hydroxyapatite nanowires. *In vitro* experiments showed that the biomedical paper possessed a good degradability, high biocompatibility, and high bioactivity.

The biomedical paper sheets with different weight percentages of ultralong hydroxyapatite nanowires were prepared by casting a suspension containing ultralong hydroxyapatite nanowires and chitosan. Figure 5.69 shows digital images of the as-prepared biomedical paper sheets with different weight percentages of ultralong hydroxyapatite nanowires from 0 wt.% to 100 wt.%. The biomedical paper sheets with different weight percentages of ultralong hydroxyapatite nanowires showed different appearances in terms of transparency, color, and surface roughness. The pure chitosan membrane was transparent and smooth on its surface with many wrinkles. As the weight percentage of ultralong hydroxyapatite nanowires

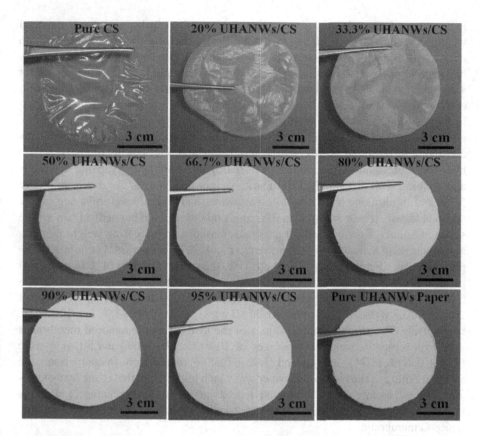

FIGURE 5.69 Digital images of the as-prepared ultralong hydroxyapatite nanowires (UHANWs)/chitosan (CS) biomedical paper sheets with different weight percentages of ultralong hydroxyapatite nanowires: 0 (pure chitosan membrane), 20 wt.%, 33.3 wt.%, 50 wt.%, 66.7 wt.%, 80 wt.%, 90 wt.%, 95 wt.%, and 100 wt.% (pure ultralong hydroxyapatite nanowire biomedical paper). (Reprinted with permission from reference [89])

increased, the biomedical paper became milk-white in color with decreased transmittance and increased roughness on the surface. Although hydroxyapatite/chitosan composite membranes had good flexibility and biological properties, the preparation of hydroxyapatite/chitosan composite membranes with high hydroxyapatite weight percentages and superior mechanical properties is still a great challenge. Kithva et al.[90] reported the preparation of a hydroxyapatite/chitosan composite film with a hydroxyapatite weight percentage of 66 wt.% using a formaldehyde-treated chitosan solution, but composite films with hydroxyapatite weight percentages of higher than 40 wt.% could not be obtained using an untreated chitosan solution. The author's research group also used hydroxyapatite nanorod membranes consisting of short hydroxyapatite nanorods (diameters of ~10 nm and lengths of ~100 nm) and chitosan with different hydroxyapatite weight percentages as control samples for comparison. The experiments indicated that hydroxyapatite nanorod membranes with enough high strength for tensile tests could not be obtained when the hydroxyapatite nanorod weight percentages were higher than 66.7 wt.%, showing that it was difficult to obtain hydroxyapatite nanorod membranes with hydroxyapatite weight percentages higher than 66.7 wt.%. In comparison, the highly flexible biomedical paper sheets with weight percentages of ultralong hydroxyapatite nanowires in a wide range from 0 wt.% to 100 wt.% could be prepared, indicating the important role of ultralong hydroxyapatite nanowires in the formation of the high-strength biomedical paper.

Figure 5.70 shows the mechanical properties of the biomedical paper sheets consisting of ultralong hydroxyapatite nanowires and chitosan in comparison with hydroxyapatite nanorod membranes consisting of short hydroxyapatite nanorods and chitosan with different hydroxyapatite weight percentages. As shown in Figure 5.70a and b, ultralong hydroxyapatite nanowires had a significant influence on the strain at failure, tensile strength, and Young's modulus of the biomedical paper. The strain at failure of the biomedical paper decreased with increasing weight percentage of ultralong hydroxyapatite nanowires (Figure 5.70c). The strain at failure of the pure chitosan membrane was 42.89 ± 6.11%, the strain at failure of the biomedical paper was (24.32 ± 2.66)%, (9.33 ± 0.47)%, (6.64 ± 0.88)%, (4.33 ± 0.22)%, (2.81 ± 0.76)%, (1.86 ± 0.14)%, (1.42 ± 0.11)%, and (0.67 ± 0.03)% for a weight percentage of 20 wt.%, 33.3 wt.%, 50 wt.%, 66.7 wt.%, 80 wt.%, 90 wt.%, 95 wt.%, and 100 wt.%, respectively. The strain at failure values of the hydroxyapatite nanorod membranes with hydroxyapatite weight percentages of 20 wt.%, 33.3 wt.%, and 50 wt.% were (7.81 ± 0.81)%, (5.54 ± 0.56)%, and (3.92 ± 0.35)%, respectively. In comparison, the strain at failure of the biomedical paper was much higher than that of the hydroxyapatite nanorod membrane with the same hydroxyapatite weight percentage, indicating that the biomedical paper possessed better toughness than that of the hydroxyapatite nanorod membrane.

The addition of ultralong hydroxyapatite nanowires had an obvious effect on the tensile strength and Young's modulus of the biomedical paper (Figure 5.70d and e). Compared with the pure chitosan membrane, the ultimate tensile strength of the biomedical paper increased with increasing weight percentage of ultralong hydroxyapatite nanowires ranging from 20 wt.% to 50 wt.%, and the tensile strength reached the maximum value of 98.93 ± 5.51 MPa at a weight percentage of ultralong hydroxyapatite nanowires of 50 wt.%; further increase of the addition amount of ultralong hydroxyapatite nanowires resulted in decrease in the tensile strength, which was 93.67 ± 3.72 MPa, 34.20 ± 7.59 MPa, 11.79 ± 1.10 MPa, 6.21 ± 0.36 MPa, and 0.39 ± 0.05 MPa, respectively, for a weight percentage of ultralong hydroxyapatite nanowires of 66.7 wt.%, 80 wt.%, 90 wt.%, 95 wt.%, and 100 wt.%, respectively. The addition of ultralong hydroxyapatite nanowires had a similar effect on the Young's modulus of the biomedical paper (Figure 5.70e). The Young's modulus of the pure chitosan membrane was 1.12 ± 0.16 GPa. The Young's modulus of the biomedical paper increased with increasing addition amount of ultralong hydroxyapatite nanowires, reached the maximum value of 3.53 ± 0.12 GPa at a hydroxyapatite nanowire weight percentage of 66.7 wt.%, which was more than three times that of the pure chitosan membrane. However, further increase of the amount of ultralong hydroxyapatite nanowires would decrease the Young's modulus of the biomedical paper.

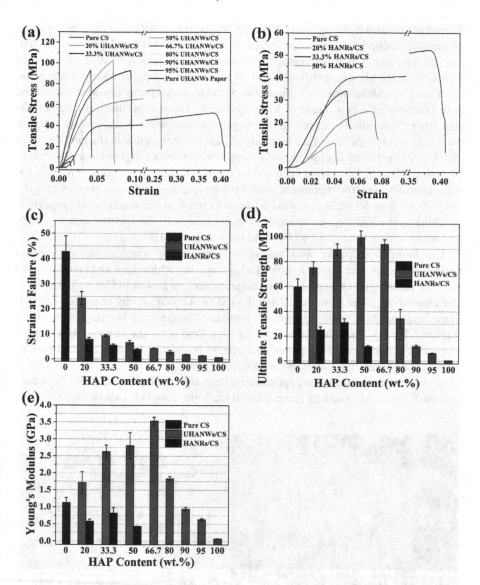

FIGURE 5.70 Mechanical properties of the biomedical paper sheets consisting of ultralong hydroxyapatite nanowires and chitosan, and hydroxyapatite nanorod membranes consisting of short hydroxyapatite nanorods (diameters of ~10 nm and lengths of ~100 nm) and chitosan with different hydroxyapatite weight percentages. (a, b) Stress-strain curves of the biomedical paper sheets (a) and hydroxyapatite nanorod membranes (b); (c) strain at failure; (d) ultimate tensile strength; (e) Young's modulus. (Reprinted with permission from reference [89])

The ultimate tensile strength and Young's modulus of the hydroxyapatite nanorod membrane reached the maximum values of 30.93 ± 3.11 MPa and 0.82 ± 0.17 GPa, respectively, at a weight percentage of hydroxyapatite nanorods of 33.3 wt.%. In comparison, the ultimate tensile strength of the biomedical paper was ~3.2 times and the Young's modulus was ~4.3 times that of the hydroxyapatite nanorod membrane at the same weight percentage of hydroxyapatite. These experimental results indicated that the addition of ultralong hydroxyapatite nanowires could greatly enhance the mechanical properties of the biomedical paper. Compared with the pure chitosan membrane, the biomedical paper with weight percentages ranging from 20 wt.% to 66.7 wt.% showed an obvious increase in the mechanical properties; however, the hydroxyapatite nanorod membrane with hydroxyapatite weight percentages ranging from 20 wt.% to 50 wt.% exhibited a great decrease in the mechanical properties. In addition, the hydroxyapatite nanorod membrane with the hydroxyapatite weight percentages higher than 66.7 wt.% could not be prepared.[89]

The biomedical paper with weight percentages of ultralong hydroxyapatite nanowires higher than 50 wt.% showed a high-quality white color and could be used for color printing using a commercial ink-jet printer, as shown in Figure 5.71a and b. The biomedical paper exhibited a high flexibility, it could rapidly restore its original shape after bending, and there was no obvious damage after bending many times (Figure 5.71c–g). The advantages such as high flexibility and superior mechanical properties of the biomedical paper will enable promising biomedical applications such as bone-fracture fixation and bone defect repair.

In order to demonstrate the application potential of the biomedical paper in bone-fracture fixation, a biomedical paper with 90 wt.% ultralong hydroxyapatite nanowires

FIGURE 5.71 Digital images of the as-prepared biomedical paper with 90 wt.% ultralong hydroxyapatite nanowires and 10 wt.% chitosan. (a) A sheet of biomedical paper with a length of 7.5 cm; (b) the biomedical paper could be used for color printing using a commercial ink-jet printer; (c–g) the biomedical paper exhibited excellent flexibility. Scale bar was 1 cm. (Reprinted with permission from reference [89])

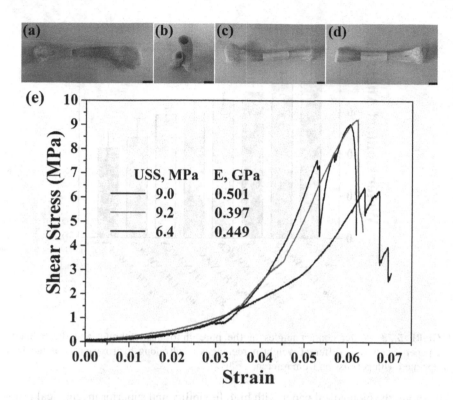

FIGURE 5.72 Digital images and mechanical properties of the biomedical paper with 90 wt.% of ultralong hydroxyapatite nanowires and 10 wt.% chitosan. Scale bar was 1 cm. (a, b) The chicken mid-femur fracture model prepared by cutting the clean femur in the middle; (c, d) the chicken mid-femur fracture model was fixed by a single layer of the biomedical paper; (e) the shear stress-shear strain curves of the chicken mid-femur fracture model fixed by a single layer of the biomedical paper. (Reprinted with permission from reference [89])

was used for investigation. A chicken mid-femur fracture was prepared by cutting the clean femur in the middle as a bone-fracture model, as shown in Figure 5.72a and b. The flexible biomedical paper with a width of ~3.0 cm was used to fix the chicken mid-femur fracture model with only one layer, and the end of the biomedical paper was stuck together with a medical biological glue (Figure 5.72c and d). The highly flexible biomedical paper could well fix the fracture model, indicating its promising application in fixing the bone fracture. The shear stress-shear strain curves of the chicken mid-femur fracture model fixed by a single layer of the biomedical paper were tested (Figure 5.72e). The shape of the shear stress-shear strain curves was different from the tensile stress-strain curves. The shear stress increased not smoothly as the shear strain increased, and did not rapidly fall at the shear strain at failure, indicating the biomedical paper fixed on the chicken mid-femur fracture model could resist the crack propagation and stably fix the fracture model under the shear force. The ultimate shear strength (USS), shear strain at failure, and Young's modulus (E) were 8.20 ± 1.56 MPa, $(6.2 \pm 0.2)\%$, and 0.45 ± 0.05 GPa, respectively.

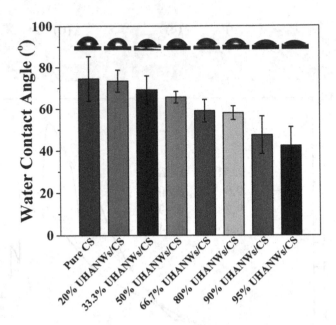

FIGURE 5.73 Water contact angles of the pure chitosan membrane and the biomedical paper sheets with different weight percentages of ultralong hydroxyapatite nanowires. (Reprinted with permission from reference [89])

Therefore, the biomedical paper with high flexibility and superior mechanical properties is promising for applications in bone-fracture fixation and bone defect repair.

The surface wettability is one of the most important parameters for the biocompatibility of biomaterials and can significantly affect the contact performance between biomaterials and tissues. As shown in Figure 5.73, the contact angles of the pure chitosan membrane and biomedical paper sheets with different weight percentages of ultralong hydroxyapatite nanowires were smaller than 80°. The contact angle of the biomedical paper decreased with increasing weight percentage of ultralong hydroxyapatite nanowires owing to their hydrophilicity originating from a large number of –OH groups on the paper surface.

The as-prepared biomedical paper is promising for biomedical applications such as wound dressing, bone-fracture fixation, artificial periosteum, and bone defect repair owing to its high flexibility, superior mechanical properties, and excellent biocompatibility. As wound dressing materials, a high swelling ratio of the biomedical paper is desirable for absorbing exudates released from the wound region and buildup of exudates to prevent excessive dehydration. In bone-fracture fixation and bone defect repair materials, a low swelling ratio of the biomedical paper is helpful to inhibit the material from severe deformation and to provide sufficient mechanical support for the new bone formation. Figure 5.74 shows the swelling ratios and water vapor transmission rates (WVTR) of the biomedical paper sheets with different weight percentages of ultralong hydroxyapatite nanowires. The pure chitosan membrane had a swelling ratio of 2.30 ± 0.43 g g^{-1}, which was higher than that of the biomedical paper. The swelling ratio of the

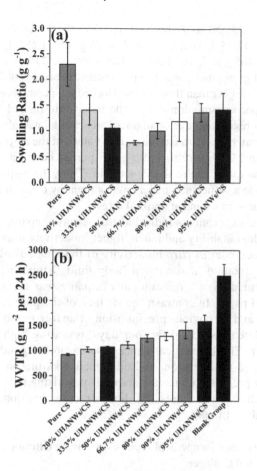

FIGURE 5.74 (a) The swelling ratio and (b) water vapor transmission rate (WVTR) of the pure chitosan membrane and the biomedical paper sheets with different weight percentages of ultralong hydroxyapatite nanowires. (Reprinted with permission from reference [89])

biomedical paper decreased and reached the minimal value of 0.77 ± 0.04 g g^{-1} at 50 wt.% ultralong hydroxyapatite nanowires. Further increase in the addition amount of ultralong hydroxyapatite nanowires resulted in increased swelling ratio of the biomedical paper. Therefore, the swelling ratio of the biomedical paper could be controlled by adjusting the addition amount of ultralong hydroxyapatite nanowires.

The water vapor transmission rates of the biomedical paper sheets with different weight percentages of ultralong hydroxyapatite nanowires are shown in Figure 5.74b. The water vapor transmission rate of the pure chitosan membrane was 922 ± 23 g m^{-2} per 24 hours. The water vapor transmission rate of the biomedical paper increased with increasing weight percentage of ultralong hydroxyapatite nanowires because of the formed porous structure in the biomedical paper which could allow the easy passing of water molecules. The water vapor transmission rates of the biomedical paper sheets with the hydroxyapatite nanowire weight percentages of 20 wt.%, 33.3 wt.%, 50 wt.%, 66.7 wt.%, 80 wt.%, 90 wt.%, and 95 wt.% were 1,028 ± 51 g m^{-2}, 1,078 ± 11 g m^{-2}, 1,121 ± 70 g m^{-2}, 1,254 ± 71 g m^{-2}, 1,293 ± 78 g m^{-2}, 1,417 ± 165 g m^{-2}, and 1,591 ± 128 g m^{-2} per 24 hours, respectively. The uncovered control group had a water vapor transmission rate of 2,493 ± 233 g m^{-2} per 24 hours, much higher than those of the biomedical paper sheets. It was reported that the water vapor transmission rate of the intact skin ranged from 240 g m^{-2} to 1,920 g m^{-2} per 24 hours, and that of an uncovered wound was about 4,800 g m^{-2} per 24 hours.[91] The water vapor transmission rate values of the as-prepared biomedical paper ranged from 1,028 g m^{-2} to 1,591 g m^{-2} per 24 hours, just within the water vapor transmission rate range of the intact skin. Thus, the as-prepared biomedical paper could provide a sufficiently moist environment by controlling the evaporative water loss from the skin wound at optimal rates.

In addition, the experiments indicated that the as-prepared biomedical paper possessed good degradability and much higher bioactivity than the pure chitosan membrane. The acellular *in vitro* bioactivity of the biomedical paper was tested by apatite mineralization in simulated body fluid. After soaking in simulated body fluid for four days, a large amount of apatite was formed on the surface of the biomedical paper. In contrast, the surface of the pure chitosan membrane was still smooth and no obvious precipitation of apatite could be observed after soaking in simulated body fluid for four days, revealing high bioactivity of the biomedical paper. The biomedical paper also exhibited good biocompatibility for cell growth and proliferation, and enhanced alkaline phosphatase activity. Therefore, the as-prepared biomedical paper is promising for various biomedical applications, such as wound dressing, bone-fracture fixation, artificial periosteum, and bone defect repair.[89]

(2) High-Performance Biomedical Paper Comprising Ultralong Hydroxyapatite Nanowires and Collagen

Collagen is an important organic constituent of biological bones, and has attracted much interest as an excellent candidate for guided bone regeneration application owing to its excellent cell affinity and high biocompatibility. Unfortunately, the mechanical properties of the pure collagen membranes are usually poor and difficult to regulate, restricting its practical application in bone regeneration. Highly flexible hydroxyapatite/collagen composite membranes are promising for application in guided bone regeneration owing to their similar chemical composition to that of natural bone, excellent biocompatibility, bioactivity, biodegradability, and osteoconductivity. However, the mechanical strengths of the hydroxyapatite/collagen composite membranes are usually low, leading to difficult surgical operations and low mechanical stability during the bone regeneration process.

The author's research group developed a new kind of highly flexible biomedical paper consisting of ultralong hydroxyapatite nanowires and collagen with a wide hydroxyapatite nanowire weight percentages ranging from 0 wt.% to 100 wt.%.[92] The as-prepared biomedical paper showed the superior mechanical properties, high flexibility, and excellent biocompatibility and cellular attachment performance, and

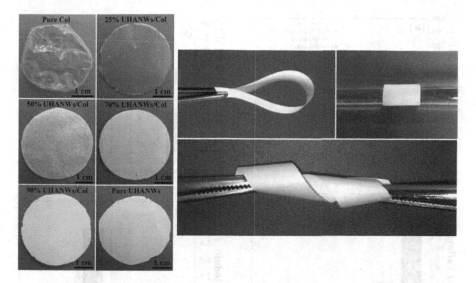

FIGURE 5.75 Digital images of the biomedical paper sheets consisting of ultralong hydroxyapatite nanowires and collagen with different weight percentages of ultralong hydroxyapatite nanowires ranging from 0 (pure collagen membrane) to 100 wt.% (pure ultralong hydroxyapatite nanowire fire-resistant inorganic paper). The biomedical paper shows high flexibility and superior mechanical properties. (Reprinted with permission from reference [92])

is promising for various biomedical applications such as guided bone regeneration and skin wound healing. The biomedical paper sheets with different weight percentages of ultralong hydroxyapatite nanowires ranging from 0 wt.% to 100wt.% (0 wt.%, 25 wt.%, 50 wt.%, 70 wt.%, 90 wt.%, and 100 wt.%) were prepared by casting the suspension containing ultralong hydroxyapatite nanowires and collagen. Figure 5.75 shows digital images of the as-prepared biomedical paper sheets consisting of ultralong hydroxyapatite nanowires and collagen with different weight percentages of ultralong hydroxyapatite nanowires ranging from 0 wt.% to 100 wt.%. The biomedical paper sheets with different weight percentages of ultralong hydroxyapatite nanowires showed different appearances in terms of transparency, color, and surface roughness. The pure collagen membrane was almost transparent and its surface is relatively smooth, but wrinkles were observed because of poor flexibility. As the weight percentage of ultralong hydroxyapatite nanowires increased, the biomedical paper showed lower transmittance, and a milk-white color. In addition, the biomedical paper showed a high flexibility, which is promising for tightly wrapping the curved bone defect surface in guided bone regeneration application.

Figure 5.76 shows mechanical properties of the biomedical paper consisting of ultralong hydroxyapatite nanowires and collagen with different weight percentages of ultralong hydroxyapatite nanowires. As shown in Figure 5.76a, the addition amount of ultralong hydroxyapatite nanowires had an obvious effect on the mechanical properties of the biomedical paper. The strain at failure of the biomedical paper decreased with increasing addition amount of ultralong

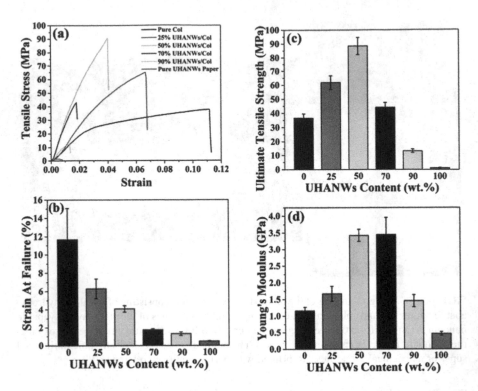

FIGURE 5.76 Mechanical properties of the biomedical paper consisting of ultralong hydroxyapatite nanowires and collagen with different weight percentages of ultralong hydroxyapatite nanowires. (a) Stress-strain curves; (b) strain at failure; (c) ultimate tensile strength; (d) Young's modulus. (Reprinted with permission from reference [92])

hydroxyapatite nanowires (Figure 5.76b). In addition, the addition amount of ultralong hydroxyapatite nanowires could affect the tensile strength and Young's modulus of the biomedical paper (Figure 5.76c and d). Compared with the pure collagen membrane, the ultimate tensile strength of the biomedical paper was significantly enhanced as the weight percentage of ultralong hydroxyapatite nanowires increased from 0 wt.% to 50 wt.%, and reached the maximum value of 88.4 ± 6.2 MPa at 50 wt.% ultralong hydroxyapatite nanowires, which was nearly 2.4 times that of the pure collagen membrane (36.5 ± 3.0 MPa). The ultimate tensile strength of the biomedical paper decreased with further increase of the weight percentage of ultralong hydroxyapatite nanowires. The Young's modulus varied in a similar way to that of the tensile strength with increasing weight percentage of ultralong hydroxyapatite nanowires (Figure 5.76d). The 50 wt.% and 70 wt.% ultralong hydroxyapatite nanowire biomedical paper sheets showed the highest Young's moduli of 3.42 ± 0.19 GPa and 3.45 ± 0.50 GPa, respectively, which were almost three times that of the pure collagen membrane (1.16 ± 0.11 GPa). In addition, the experiments indicated that the as-prepared highly flexible 70 wt.% ultralong hydroxyapatite nanowire/collagen biomedical paper possessed

a good degradability, excellent biocompatibility, and excellent cellular attachment performance compared with the pure collagen membrane and 70 wt.%
hydroxyapatite nanorod/collagen membrane. Considering the high biocompatibility, superior mechanical properties, and excellent cellular attachment performance, the as-prepared biomedical paper is promising for various applications in
biomedical fields, such as guided bone regeneration and skin wound healing.[92]

5.7.3. BONE DEFECT REPAIR AND BONE REGENERATION

Bone is a mineralized tissue formed with a composition of 70 wt.% minerals (mostly
hydroxyapatite nanocrystals) and 30 wt.% organics (mainly collagen). It is well
known that the extracellular matrix in bone is an assembled network of mineralized
collagen fibrils (collagen and hydroxyapatite nanocrystals). Ordered microstructures
with a preferred orientation of mineralized collagen fibrils enable mechanical properties of bone with an extraordinary combination of high strength and toughness.
One-dimensional hydroxyapatite nanowires can mimic the structure of mineralized
collagen fibrils of bone and exhibit superior mechanical properties.

The guided bone regeneration technique, a well-established therapy, is considered
one of the most promising techniques for bone defect repair and bone regeneration
owing to its high efficiency and simplicity. In the guided bone regeneration technique, a barrier membrane is needed to cover around the bone defect area and inhibit
the invasion of the fibrous connective tissue, thus promoting the growth of cells
derived from the bone marrow and the formation of new bone in the bone defect
area. Non-bioresorbable and bioresorbable membranes are two main types of barrier membranes. The nonbioresorbable membranes used as barriers for guided bone
regeneration are bio-inert and need to be removed by a secondary surgical operation, which increases the risk of undesirable side effects. The bioresorbable barrier
membranes are promising for application in guided bone regeneration because of
their high biodegradability and as they do not need a second surgery. Unfortunately,
the degradation of some bioresorbable membranes such as poly(lactic acid) and
poly(lactic-co-glycolic acid) can produce acids during the resorption process, resulting in an inflammatory response.[92]

As discussed above, the biomedical paper based on ultralong hydroxyapatite
nanowires and biopolymers have many advantages, such as high biocompatibility,
good biodegradability, high flexibility, superior mechanical properties, and porous
structure. Therefore, the biomedical paper based on ultralong hydroxyapatite nanowires are promising for various biomedical applications, such as bone defect repair,
skin wound healing, drug/protein/growth factor delivery, and rapid testing paper. It
is expected that the biomedical paper based on ultralong hydroxyapatite nanowires
has promising application as the high-performance artificial periosteum and barrier
membrane in guided bone regeneration.

Studies were performed to investigate the bone defect repair performance of the
highly flexible biocompatible inorganic biomedical paper consisting of ultralong
hydroxyapatite nanowires as the only building material without any organic constituent.[93] Circular-shaped cranial bone defects with a diameter of 8.8 mm were

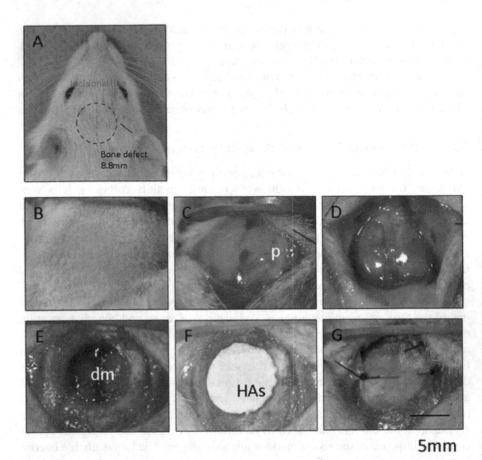

FIGURE 5.77 Digital images of surgical procedures for making an animal cranial bone defect model. (A) A rat head indicating lines of a median sagittal incision and a bone defect; (B) the shaving condition of surgical area; (C) the incision to expose the periosteum; (D) the exposure of the cranial bone by peeling the periosteum; (E) the removal of the parietal bone from the cerebral dura mater using a trephine bar; (F) an inorganic biomedical paper was placed in the bone defect area; (G) the suturing of the periosteum and the skin. (Reprinted with permission from reference [93])

prepared to expose the dura mater in Wistar rats, as shown in Figure 5.77. A similar-sized, circular-shaped ultralong hydroxyapatite nanowire biomedical paper was placed in the bone defect area.

After two, four, and eight weeks post-implantation, the rats were sacrificed, and the experimental sections were examined by micro-CT scanning and histological observation. The ultralong hydroxyapatite nanowire biomedical paper showed no inflammatory effect, and the paper thickness decreased over time with a good biodegradation performance. The experiments indicated that tartrate-resistant acid phosphatase-positive osteoclast-like cells were induced around the edges of the ultralong hydroxyapatite nanowire biomedical paper. The newly formed bone was found in the bone defect area, either with a direct contact with the ultralong hydroxyapatite nanowire biomedical paper or through thin fibrous tissues. The ultralong hydroxyapatite nanowire biomedical paper-induced osteoblast differentiation was found since the alkaline phosphatase activities were detected on the surface of the ultralong hydroxyapatite nanowire biomedical paper. These results indicated that the ultralong hydroxyapatite nanowire biomedical paper could induce osteogenesis without causing any harmful effects.

The experiments showed that the new fibrous connective tissues were formed on the surface of the ultralong hydroxyapatite nanowire biomedical paper facing the skin. At week 2, the granulation tissues with abundant capillaries were observed in the ultralong hydroxyapatite nanowire biomedical paper group. A high degree of inflammatory cell infiltration, which generally happened in response to infection and/or harmful foreign materials, was not observed. Elastica-Masson-stained specimens exhibited layers consisting of newly formed fibrous connective tissues. At week 4, tissues surrounding the ultralong hydroxyapatite nanowire biomedical paper were replaced by fibrous connective tissues. The thicknesses of fibrous connective tissues (called fibrous capsules) surrounding the ultralong hydroxyapatite nanowire biomedical paper decreased at week 4. The surrounding tissues consisted of fibrous components including dense collagen fibers and cellular components. At week 8, cellular components were reduced, and high-density collagen fibers were observed. Both the thicknesses of fibrous capsules in the control (blank) group and ultralong hydroxyapatite nanowire biomedical paper group significantly decreased from week 2 to week 4 or week 8, and the thicknesses of fibrous capsules in the ultralong hydroxyapatite nanowire biomedical paper group were significantly thinner than those in the control group.

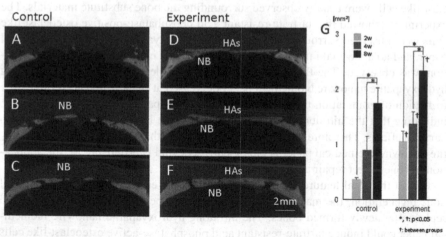

FIGURE 5.78 Micro-CT images in the bone defect area in the control group (A, B, and C) and the ultralong hydroxyapatite nanowire biomedical paper group (D, E, and F). (A, D) 2 weeks; (B, E) 4 weeks; (C, F) 8 weeks (HAs stands for the ultralong hydroxyapatite nanowire biomedical paper; NB stands for newly formed bone); (G) the amounts of new bone formation. (Reprinted with permission from reference [93])

Figure 5.78 shows micro-CT images in the bone defect area in the control group and the ultralong hydroxyapatite nanowire biomedical paper group. New bone formation was observed in the periphery of the bone defect area facing the dura mater. At week 2, new bones formed directly below the ultralong hydroxyapatite nanowire biomedical paper, while almost no new bone was observed in the control group (Figure 5.78A and D). At week 8, in both the control group and the ultralong hydroxyapatite nanowire biomedical paper group, new bones were observed in the central part of the bone defect area. Histologically, new bone formation gradually increased over time (Figure 5.78E-F). Micro-CT images showed significantly increased new bone formation in both the control group and the ultralong hydroxyapatite nanowire biomedical paper group with increasing time (Figure 5.78G), and the amount of newly formed bone was obviously higher in the ultralong hydroxyapatite nanowire biomedical paper group than in the control group.[93]

Although hydroxyapatite biomaterials have excellent biocompatibility, they have shortcomings; for example, their mechanical properties such as rigidity, brittleness, and hardness cannot match with the softness of living tissues. In contrast, ultralong hydroxyapatite nanowires have ultrathin diameters of approximately 10 nm, ultralong lengths up to several hundred micrometers and ultrahigh aspect ratios up to more than 10,000. Highly flexible ultralong hydroxyapatite nanowires can self-assemble into bundles and be interwoven with each other to form the highly flexible biomedical paper with a porous layered structure that is desirable for bone regeneration.

In a living organism, foreign substances will induce macrophage responses. Even biocompatible materials such as poly(lactic acid) can cause macrophage activation when they are degraded into small debris, leading to the formation of granulation tissues in a long period of time. Tartrate-resistant acid phosphatase-positive osteoclast-like cells were usually observed surrounding the bone substitute materials. The experiments showed that tartrate-resistant acid phosphatase-positive osteoclast-like cells were observed surrounding the ultralong hydroxyapatite nanowire biomedical paper, and no abnormal inflammatory cell infiltration, including macrophage induction, was observed. The thicknesses of fibrous capsules surrounding the ultralong hydroxyapatite nanowire biomedical paper decreased at week 8, and accordingly the formation of granulation tissues was suppressed compared with the control group, indicating that the ultralong hydroxyapatite nanowire biomedical paper caused no harmful effects. Therefore, the experiments indicated that the ultralong hydroxyapatite nanowire biomedical paper is biologically safe and is promising for the application in bone defect repair and bone regeneration.[93]

One of the ideal features for bone substitution materials is that after the guided bone regeneration, the materials are degraded, resorbed in a living organism, and replaced by newly formed bones. The ultralong hydroxyapatite nanowire biomedical paper could induce tartrate-resistant acid phosphatase-active osteoclast-like cells especially at the edge of the paper, suggesting that the ultralong hydroxyapatite nanowire biomedical paper was degraded and resorbed from the paper edge and slowly replaced by newly formed bones. The previous study by the author's research group revealed that the flexible biomedical paper based on ultralong hydroxyapatite nanowires could slowly degrade in the normal saline solution *in vitro*.[89, 92] The cell activity usually depends on the surface composition and morphology of the materials to which the cells attach. Histological findings showed the alkaline phosphatase activities in bone cells that were in direct contact with the ultralong hydroxyapatite nanowire biomedical paper, indicating that the ultralong hydroxyapatite nanowire biomedical paper could induce the osteoblast differentiation that was also observed in the layer-structured interspaces. Based on the unique merits of the ultralong hydroxyapatite nanowire biomedical paper, it is a highly potential candidate for establishing novel bone therapies; for example, covering a large-sized bone defect area with the ultralong hydroxyapatite nanowire biomedical paper after filling the defect with a conventional bone substitute may prohibit soft tissues from invading to the bone defect area, and at the same time, promote the new bone formation.[93] In addition, the ultralong hydroxyapatite nanowire biomedical paper may be used as a high-performance artificial periosteum for bone defect repair.[88]

In another research work, a highly flexible biomedical paper consisting of selenium-doped ultralong hydroxyapatite nanowires and chitosan was investigated for high-performance anti-bone tumor application.[94] Selenium is one of the essential elements for human health with important functions for the prevention of cardiovascular diseases, cognitive disorder, and viral infection. In addition, selenium was investigated for the application in the therapeutic treatment of the cancer. Current clinical treatment of osteosarcoma is usually directed resection, leading to side effects. The as-prepared biomedical paper consisting of selenium-doped ultralong hydroxyapatite nanowires and chitosan could effectively inhibit the growth of bone tumor. The possible mechanisms of the anti-tumor effect of the biomedical paper were investigated in terms of reactive oxygen species accumulation and the activation of apoptosis and the underlying signal pathway involved. Furthermore, *in vivo* evaluations were carried out by a patient-derived xenograft animal model, and the experimental results further revealed the obvious anti-tumor effects of the biomedical paper consisting of selenium-doped ultralong hydroxyapatite nanowires and chitosan.

The effect of the biomedical paper sheets consisting of selenium-doped ultralong hydroxyapatite nanowires and chitosan with different selenium contents on the viability of HCS-2/8 chondrosarcoma cells and SJSA osteosarcoma cells was evaluated. The experiments showed that the biomedical paper possessed a superior performance in apoptosis induction and viability inhibition for both HCS-2/8 chondrosarcoma cells and SJSA osteosarcoma cells. The biomedical paper with a selenium doping molar ratio (3%) exhibited a much better apoptosis induction compared with the blank control and the biomedical paper with lower selenium doping ratios.

The biomedical paper sheets consisting of selenium-doped ultralong hydroxyapatite nanowires and chitosan could induce higher apoptosis and necrosis rate for HCS-2/8 chondrosarcoma cells and SJSA osteosarcoma cells than the blank control and chitosan sample, as shown in Figure 5.79a–d. In addition, the increased expression of cleaved caspase-3 and caspase-9 was found when the cells were co-cultured with the biomedical paper (Figure 5.79e). The activation of initiated cleaved caspase-9 could activate the caspase-3 into the cleaved caspase-3 to induce apoptosis. Furthermore, the expression of the anti-apoptotic protein Bcl-2 in the cells decreased after

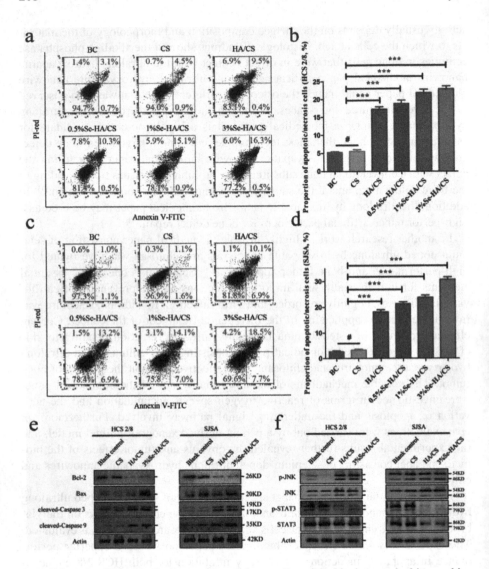

FIGURE 5.79 The biological properties of the samples of the blank control, chitosan, biomedical paper sheets consisting of selenium-doped or undoped ultralong hydroxyapatite nanowires and chitosan. (a, b) Apoptosis/necrosis proportion of HCS 2/8 chondrosarcoma cells co-cultured with different samples; (c, d) SJSA osteosarcoma cells co-cultured with different samples; (e) expression of cleaved caspase-3, caspase-9, Bax and Bcl-2 in HCS 2/8 chondrosarcoma and SJSA osteosarcoma cells co-cultured with different samples; (f) expression of JNK, p-JNK, STAT3 and p-STAT3 in HCS 2/8 chondrosarcoma and SJSA osteosarcoma cells co-cultured with different samples. (Reprinted with permission from reference [94])

co-culture with the biomedical paper, while the pro-apoptotic protein Bax increased accordingly (Figure 5.79f). These experimental results indicated the biomedical paper could induce the apoptosis process. The experiments also indicated that the tumor inhibition effect of the biomedical paper was related with a higher expression of JNK and lower expression of STAT3, especially the phosphatized forms, and could enhance the phosphorylation level of JNK and accordingly decrease the STAT3 phosphorylation in the SJSA osteosarcoma cells.

The experiments showed a higher level of reactive oxygen species in cells after co-culture with the biomedical paper consisting of selenium-doped ultralong hydroxyapatite nanowires and chitosan. The excessive reactive oxygen species could disorganize the cellular oxidation level and induce DNA damage and apoptosis in cancer cells.

In vivo anti-tumor tests were performed using the biomedical paper sheets consisting of selenium-doped or undoped ultralong hydroxyapatite nanowires and chitosan, and control samples. The patient-derived xenograft model was adopted to simulate the residual lesions after the tumor resection to investigate the in vivo inhibition effect of the biomedical paper. The patient-derived xenograft model was usually performed by subcutaneous implantation of fresh tumor tissue into immunodeficient mice, which could represent morphological and genetic features of the derived tissue and thus could be used as a preclinical model to evaluate the therapeutic effect for bone tumors. Compared with the traditional cell line-based xenografts, the patient-derived xenograft model has more similarities with the primary tumors, as shown in Figure 5.80a. Figure 5.80b and c shows that the volume of implanted patient-derived xenograft tumor in the selenium-doped biomedical paper group was the smallest among the four groups, but the blank control group exhibited the largest tumor volume. The tumor volume in the pure chitosan membrane group was slightly smaller than that in the blank control group, indicating that the isolation effect of the implanted biomedical paper led to the relative malnutrition and smaller size of the tumor. However, this isolation effect was not efficient for inhibiting the tumor growth. In comparison, the groups of the undoped and selenium-doped biomedical paper exhibited a more obvious inhibition effect on tumor growth, especially for the group of selenium-doped biomedical paper. In addition, the tumor weight was obviously smaller in the selenium-doped biomedical paper group compared with the blank control and other groups (Figure 5.80d). Figure 5.80e shows the images of tumor tissue slices with hematoxylin-eosin, proliferating cell nuclear antigen, and cleaved caspase-3 staining, which were detached from the blank control, chitosan, and undoped and selenium-doped biomedical paper groups. The hematoxylin-eosin staining images showed that the tumor tissues in the selenium-doped biomedical paper group were the most incompact compared with other groups. Furthermore, the immunohistochemical stainings of proliferating cell nuclear antigen and cleaved caspase-3 were applied for the analysis of the anti-tumor mechanisms. The immunohistochemistry assay indicated the slowest growth and enhanced apoptosis of tumors in the selenium-doped biomedical paper group compared with other groups. The selenium-doped biomedical paper group showed the lowest expression of proliferating

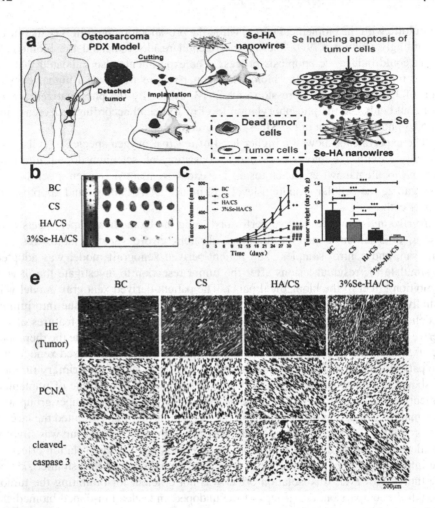

FIGURE 5.80 (a) The schematic illustration of the *in vivo* tumor inhibition induced by the biomedical paper consisting of selenium-doped ultralong hydroxyapatite nanowires and chitosan; (b–d) the general observation of patient-derived xenograft at day 30 (b), tumor volume (c), and tumor weight at day 30 (d) showed a consistent anti-cancer effect of the undoped and selenium-doped biomedical paper sheets; (e) histological images demonstrated the inactivation of PCNA and activation of cleaved caspase-3. (Reprinted with permission from reference [94])

cell nuclear antigen and highest expression of cleaved caspase-3, indicating the best anti-tumor performance among the four groups.[94]

5.7.4. RAPID TEST PAPER

In recent years, rapid analysis and detection technologies have been widely used in various fields such as chemical detection, medical diagnosis, judicial identification, environmental monitoring, and food detection. Several decades ago, the instrumental analysis and testing work were often performed by certain specific departments such as scientific research institutions, hospitals, and analytical testing centers. The instrumental analysis method has high measurement accuracy and low detection limit, but because the instruments used are generally large precision equipment, and alternating current power is used as the power source, the operation is more complicated and inconvenient to use, and it is generally not suitable for on-site rapid detection. With the advancement of science and technology, various on-site, temporary, fast, and efficient analysis and detection methods have appeared one after another. Many rapid on-site analysis and detection methods are achieved through the color change and the degree of change.

As a fast on-site detection method, the test paper method is characterized by advantages such as simple operation, convenient carrying, low price, and good selectivity, accuracy, and sensitivity, and it has wide applications in the tests of medical and health, food, water quality, air, etc. For example, the early pregnancy test paper on the market can provide a fast and efficient detection method for women to determine whether they are pregnant at an early stage. However, natural enzymes such as glucose oxidase and catalase are expensive, and their preparation, purification, and storage are time-consuming, and their activities are easily affected by the external environment such as the pH value and temperature. In contrast, artificial mimic enzymes synthesized by chemical methods have low cost and relatively stable catalytic activity, and are expected to replace some natural enzymes in the field of analysis and detection. In particular, nanostructured materials–based artificial mimic enzymes have been extensively investigated because of their tunable structures and chemical compositions, and superior properties. However, many studies focused on powdered nanostructured materials–based artificial mimic enzymes, but nanoscale powders were difficult to recover for reuse. It is expected that the flexible rapid test paper will have advantages such as simplicity, low cost, easy storage and transportation, and excellent recyclability. Ultralong hydroxyapatite nanowires are highly flexible, biocompatible, environmentally friendly, and highly efficient for absorbing organic compounds; thus, they are promising for applications in constructing various kinds of the rapid test paper.

The ultralong hydroxyapatite nanowire/metal-organic framework enzyme-mimetic core/shell nanofibers with a peroxidase-like activity were synthesized by a template method.[95] Ultralong hydroxyapatite nanowires were adopted as a hard template for the nucleation and growth of MIL-100(Fe) (a typical metal-organic framework) by a layer-by-layer deposition strategy. The Coulombic and chelation interactions between Ca^{2+} ions on the surface of ultralong hydroxyapatite nanowires and the $-COO^-$ organic linkers of MIL-100(Fe) played an important role in the formation process. The as-prepared water-stable enzyme-mimetic core/shell nanofibers possessed a peroxidase-like activity toward the oxidation of different peroxidase substrates in the presence of hydrogen peroxide, accompanied by a color change of the solution. In addition, a flexible, recyclable rapid test paper was prepared by using enzyme-mimetic core/shell nanofibers consisting of ultralong hydroxyapatite nanowires and MIL-100(Fe) as building blocks by vacuum-assisted filtration. A simple, low-cost, and sensitive colorimetric method was developed for the detection of hydrogen peroxide, glucose, and ascorbic acid based on the as-prepared rapid test paper. In addition, the as-prepared rapid test paper could be recovered easily for reuse by simply dipping in absolute ethanol for just 30 minutes, exhibiting excellent recyclability. With its combination of advantages such as high biocompatibility, easy transportation, easy storage and use, rapid recyclability, light weight, and high flexibility, the as-prepared rapid test paper is promising for wide applications in various fields.

The peroxidase-like activity of enzyme-mimetic core/shell nanofibers consisting of ultralong hydroxyapatite nanowires and MIL-100(Fe) was investigated by the catalytic oxidation of peroxidase substrate 3,3',5,5'-tetramethylbenzidine in the presence of H_2O_2. As shown in Figure 5.81a, no clear oxidation reaction occurred without the enzyme-mimetic core/shell nanofibers as the catalyst in the absence of H_2O_2, as indicated by the colorless 3,3',5,5'-tetramethylbenzidine solution. Similarly, no obvious color change was observed in the 3,3',5,5'-tetramethylbenzidine solution without enzyme-mimetic core/shell nanofibers in the presence of H_2O_2. Ultralong hydroxyapatite nanowires or enzyme-mimetic core/shell nanofibers alone could not induce the color change. Additional control experiments were conducted using ultralong hydroxyapatite nanowires or enzyme-mimetic core/shell nanofibers in the absence of H_2O_2 but in the presence of 3,3',5,5'-tetramethylbenzidine, and negligible color changes were observed. However, the addition of enzyme-mimetic core/shell nanofibers into the solution containing 3,3',5,5'-tetramethylbenzidine and H_2O_2 induced a dark green color, but no obvious color change was observed in the case of ultralong hydroxyapatite nanowires. The experiments indicated that enzyme-mimetic core/shell nanofibers had intrinsic peroxidase-like activity, and could catalyze the oxidation of the peroxidase substrate 3,3',5,5'-tetramethylbenzidine in the presence of H_2O_2, as shown in Figure 5.81b. Fe^{3+}/Fe^{2+} ions (Fenton's reagent) could act as active centers and break down H_2O_2 molecules into OH radicals. The charge-transfer complexes could form from the one-electron oxidation of 3,3',5,5'-tetramethylbenzidine by −OH radicals, forming two intense characteristic absorption peaks at ~370 nm and ~658 nm. The absorbance at 658 nm was monitored to evaluate the optimal catalytic reaction conditions. The activity of enzyme-mimetic core/shell nanofibers was dependent on the temperature, pH value, and H_2O_2 concentration. The experiments indicated that the H_2O_2 concentration could affect the catalytic activity of enzyme-mimetic core/shell nanofibers. Furthermore, the catalytic activity of enzyme-mimetic core/shell nanofibers could be adjusted by different layer-by layer deposition cycles used for the preparation of enzyme-mimetic core/shell nanofibers (Figure 5.81c). The larger the number of layer-by layer deposition cycles, the higher the content of metal-organic framework, and the higher the catalytic activity of enzyme-mimetic core/shell nanofibers.[95]

Quantitative detection of H_2O_2, glucose, and ascorbic acid was demonstrated using enzyme-mimetic core/shell nanofibers consisting of ultralong hydroxyapatite nanowires and MIL-100(Fe). A colorimetric method for the quantitative detection of H_2O_2 using the peroxidase-like activity of the enzyme-mimetic core/shell nanofibers in the presence of 3,3',5,5'-tetramethylbenzidine was developed, as shown in Figure 5.82.

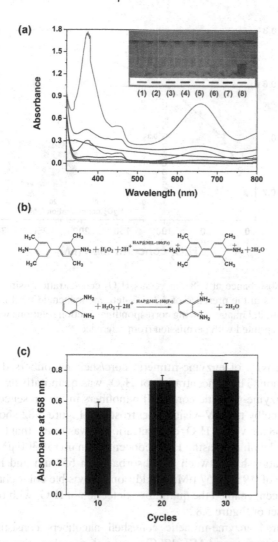

FIGURE 5.81 (a) UV–visible spectra of various CH_3COOH/CH_3COONa buffer solutions (pH 4.0) containing different constituents at 40°C for 20 minutes: (1) 3,3',5,5'-tetramethyl-benzidine; (2) 3,3',5,5'-tetramethylbenzidine + H_2O_2; (3) ultralong hydroxyapatite nanowires; (4) enzyme-mimetic core/shell nanofibers; (5) 3,3',5,5'-tetramethylbenzidine + ultralong hydroxyapatite nanowires; (6) 3,3',5,5'-tetramethylbenzidine + enzyme-mimetic core/shell nanofibers; (7) 3,3',5,5'-tetramethylbenzidine + H_2O_2 + ultralong hydroxyapatite nanowires; (8) 3,3',5,5'-tetramethylbenzidine + H_2O_2 + enzyme-mimetic core/shell nanofibers. Inset: a digital image of the corresponding solutions; (b) the reaction equations for the catalytic oxi-dation of peroxidase substrates: 3,3',5,5'-tetramethylbenzidine (top) and o-phenylenediamine (bottom) in the presence of H_2O_2 and enzyme-mimetic core/shell nanofibers; (c) analysis of tunable catalytic activity of different enzyme-mimetic core/shell nanofibers of hydroxyapatite nanowire/MIL-100(Fe)-10, -20, and -30 prepared using different layer-by-layer deposition cycles by monitoring the absorbance at 658 nm. Concentrations: [3,3',5,5'-tetramethylbenzi-dine] = 0.19 mM, $[H_2O_2]$ = 38 mM, [enzyme-mimetic core/shell nanofibers] = 38 μg mL^{-1}. (Reprinted with permission from reference [95])

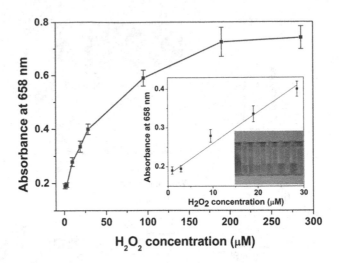

FIGURE 5.82 Absorbance at 658 nm versus H_2O_2 concentration using enzyme-mimetic core/shell nanofibers in the presence of 3,3',5,5'-tetramethylbenzidine. Insets: a linear calibration plot and a digital image showing corresponding reaction solutions with different H_2O_2 concentrations. (Reprinted with permission from reference [95])

The catalytic activity of enzyme-mimetic core/shell nanofibers depended on the H_2O_2 concentration. The concentration of H_2O_2 was quantitatively evaluated using the powdered enzyme-mimetic core/shell nanofibers in the presence of 3,3',5,5'-tetramethylbenzidine by the UV–visible spectroscopy. Figure 5.82 shows the curve of absorbance at 658 nm versus H_2O_2 concentration. It was found that the absorbance at 658 nm increased with increasing H_2O_2 concentration up to ~300 μM, and that there was a linear relationship between the absorbance at 658 nm and H_2O_2 concentration in the range of 0.95–28.57 μM. In addition, the visible color change (from light green to deep green) enabled the qualitative detection of H_2O_2 with the naked eye, as shown in the inset of Figure 5.82.

The as-prepared enzyme-mimetic core/shell nanofibers consisting of ultralong hydroxyapatite nanowires and MIL-100(Fe) were also used for the quantitative detection of glucose in the presence of 3,3',5,5'-tetramethylbenzidine. Glucose is an important analyte in the food industry and clinical diagnostics. As shown in Figure 5.83a, glucose could be oxidized to form gluconic acid and H_2O_2 in the presence of glucose oxidase (GOx) and oxygen. Based on this phenomenon, a colorimetric method was designed for the quantitative detection of glucose using enzyme-mimetic core/shell nanofibers in the presence of 3,3',5,5'-tetramethylbenzidine. Figure 5.83b shows the curve of absorbance at 658 nm versus glucose concentration. A linear relationship was found between the absorbance at 658 nm and glucose concentrations ranging from 2 μM to 50 μM. Furthermore, the color change of solutions with various glucose concentrations was visible with the naked eye (the inset of Figure 5.83b). The specificity of the colorimetric method for glucose detection was also evaluated. The catalytic reaction was performed in the presence of fructose, lactose, or maltose instead of glucose using enzyme-mimetic

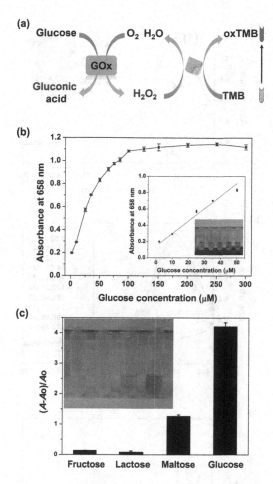

FIGURE 5.83 (a) Schematic illustration of a colorimetric method for glucose detection using enzyme-mimetic core/shell nanofibers consisting of ultralong hydroxyapatite nanowires and MIL-100(Fe) in the presence of 3,3',5,5'-tetramethylbenzidine; (b) absorbance at 658 nm versus glucose concentration using enzyme-mimetic core/shell nanofibers in the presence of 3,3',5,5'-tetramethylbenzidine. Insets: a linear calibration plot and a digital image showing corresponding reaction solutions with different glucose concentrations; (c) evaluation of the selectivity for glucose detection by monitoring the absorbance at 658 nm. Analyte concentration was 100 μM for fructose, lactose, maltose, or glucose. Inset: a digital image showing reaction solutions. (A0 and A are the absorbance of the blank solution and the reaction solution, respectively, at 658 nm). (Reprinted with permission from reference [95])

core/shell nanofibers in the presence of 3,3',5,5'-tetramethylbenzidine. The significant difference in the absorbance and solution color for glucose compared with fructose, lactose, and maltose indicated a high selectivity for glucose of this method (Figure 5.83c).

Another peroxidase substrate, o-phenylenediamine, could also be oxidized catalytically by enzyme-mimetic core/shell nanofibers consisting of ultralong

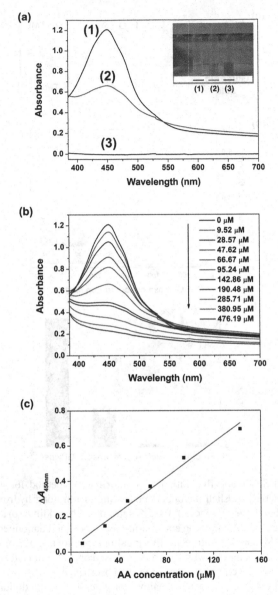

FIGURE 5.84 (a) UV–visible spectra of various CH_3COOH/CH_3COONa buffer solutions (pH 4.0) containing different constituents at room temperature for 15 minutes: (1) o-phenylenediamine; (2) o-phenylenediamine + H_2O_2 + enzyme-mimetic core/shell nanofibers + ascorbic acid; (3) o-phenylenediamine + H_2O_2 + enzyme-mimetic core/shell nanofibers. Inset: a digital image showing corresponding reaction solutions. Concentrations: [o-phenylenediamine] = 2.38 mM; [H_2O_2] = 0.23 M; [enzyme-mimetic core/shell nanofibers] = 38 μg mL^{-1}; [ascorbic acid] = 95.24 μM; (b) UV–visible spectra of o-phenylenediamine solutions in the presence of enzyme-mimetic core/shell nanofibers, H_2O_2, and ascorbic acid with different concentrations; (c) a linear calibration plot of the ascorbic acid concentration dependent on absorbance at 450 nm. (Reprinted with permission from reference [95])

hydroxyapatite nanowires and MIL-100(Fe) in the presence of H_2O_2 to develop a yellow color and an intense characteristic peak at 450 nm, as shown in Figure 5.84a. The introduction of a trace amount of ascorbic acid, known as vitamin C and an antioxidant, could inhibit the catalytic oxidation reaction, and thus, the yellow color faded (Figure 5.84a). Thus, an indirect quantitative detection of ascorbic acid was developed using enzyme-mimetic core/shell nanofibers. The curve of absorbance at 450 nm versus ascorbic acid concentration (Figure 5.84b) shows that the peak intensity at 450 nm decreased with increasing ascorbic acid concentration. A linear relationship between absorbance at 450 nm and ascorbic acid concentration was found in the ascorbic acid concentration range from 9.52 µM to 142.86 µM, as shown in Figure 5.84c.[95]

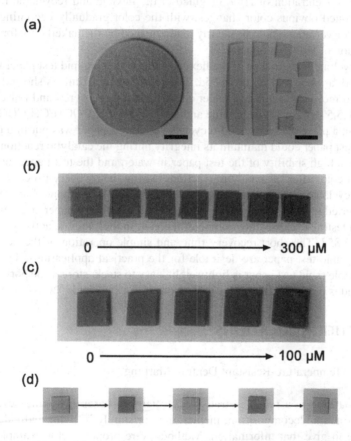

FIGURE 5.85   (a) Flexible recyclable rapid test paper sheets with different sizes and shapes based on enzyme-mimetic core/shell nanofibers consisting of ultralong hydroxyapatite nanowires and MIL-100(Fe); scale bar = 1 cm; (b, c) digital images showing the colorimetric detection of H_2O_2 with different concentrations (0 µM, 25 µM, 50 µM, 75 µM, 100 µM, 200 µM, and 300 µM) (b), and glucose with different concentrations (0, 25, 50, 75, and 100 µM) (c) using flexible recyclable rapid test paper; (d) sequential testing cycles showing excellent recyclability of the flexible recyclable rapid test paper. (Reprinted with permission from reference [95])

Qualitative detection of H_2O_2 and glucose was also investigated using the flexible and recyclable rapid test paper based on enzyme-mimetic core/shell nanofibers consisting of ultralong hydroxyapatite nanowires and MIL-100(Fe), as shown in Figure 5.85. The flexible and recyclable rapid test paper could be cut into pieces of various sizes and shapes for specific applications (Figure 5.85a). A colorimetric method for the qualitative detection of H_2O_2 and glucose was designed using the flexible and recyclable rapid test paper. During the catalytic reaction process, the high specific surface area and porous network of the flexible and recyclable rapid test paper could facilitate the diffusion of molecules and ions into the high-density active sites inside the porous framework. Figure 5.85b and c shows the qualitative detection of H_2O_2 and glucose using the flexible and recyclable rapid test paper, respectively. With increasing concentration of H_2O_2 or glucose, the flexible and recyclable rapid test paper exhibited obvious color changes, with the color gradually becoming darker, and the color variation could be clearly distinguished by the naked eyes for qualitative detection of H_2O_2 and glucose.

The recycling performance of the flexible and recyclable rapid test paper was also investigated, as shown in Figure 5.85d. The recycling experiments showed that the flexible and recyclable rapid test paper could be easily recovered and reused. After adding 3,3',5,5'-tetramethylbenzidine and H_2O_2 in the CH_3COOH/CH_3COONa buffer solution, a piece of flexible and recyclable rapid test paper was put into the solution. The test paper could maintain its integrity during the catalytic reaction process owing to the high stability of the test paper in water, and the test paper turned to a dark green color after the catalytic reaction. During the recovery process, the flexible and recyclable rapid test paper was immersed in absolute ethanol for 30 minutes, then it reverted to its original yellow color. The recovered test paper could be reused in the next testing cycles without obvious decrease in the catalytic activity, as shown in Figure 5.85d. The short recovery time and simple operation of the flexible and recyclable rapid test paper are desirable for the practical applications. The flexible and recyclable rapid test paper is lightweight, easy to stack, store, transport, use, and recycle, and is promising for wide applications in various fields.[95]

5.8.   OTHER APPLICATIONS

(1) High-Temperature-Resistant Denture Marking

Denture marking is a reliable strategy especially in cases involving severe head and neck trauma, decomposition, incineration, or simply in living individuals who are unable to give that information. Methods were proposed, for example, marks were carved or painted on the surface of the dentures, generally acrylic, and the identifying elements were inserted inside them. Thermally stable markings are very important in the case of air travel accidents where the temperatures can reach about 900–1100°C. Nowadays, quick response (QR) codes are one of the fast and efficient technologies for linking and accessing the identifying information, and their application in forensic medicine is promising. In an emergency, rapid communication is required, where the access to the contact information is very important. In such cases, the mobile telephones are convenient for the reading and interpretation of data. Today, the QR codes represent a promising technology for the identification and handling of information as they can code a hyperlink, so that reading the code allows any mobile reading device to access specific sites where the data are provided without the need for typing.

In a recent research work, the high-temperature-resistant performances of four types of paper with printed QR codes were investigated: (I) the common cellulose fiber paper, (II) the glass fiber film, (III) the ultralong hydroxyapatite nanowire paper, and (IV) the polyolefin/silica membrane, all samples were exposed to temperatures between 100°C and 1,000°C for 1 hour. Each sample was scanned using three different smart phones, and the scans were positive for (I) 33.33%, (II) 50%, (III) 100%, and (IV) 70.37%, as shown in Figure 5.86. Printed QR codes on the ultralong hydroxyapatite nanowire fire-resistant paper showed a great potential for safely preserving identifying information and QR codes even under high-temperature conditions.[96]

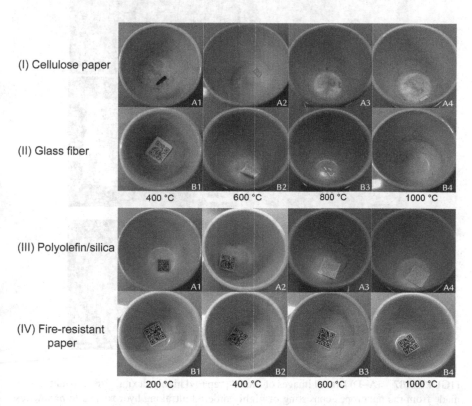

FIGURE 5.86 High-temperature-resistant performances of four types of paper with printed QR codes: (I) the common cellulose fiber paper; (II) the glass fiber film; (III) the ultralong hydroxyapatite nanowire fire-resistant paper; and (IV) the polyolefin/silica membrane. All samples were exposed to temperatures between 100°C and 1,000°C for 1 hour. (Reprinted with permission from reference [96])

(2) Highly Flexible Fire-Resistant Textiles

As discussed in Section 2.5, the highly flexible nanorope consisting of highly ordered ultralong hydroxyapatite nanowires could be prepared using the original solvothermal mother slurry containing ultralong hydroxyapatite nanowire by a simple injection into absolute ethanol.[35] Figure 5.87 shows the as-prepared highly flexible fire-resistant textiles made from the nanoropes consisting of highly ordered ultralong hydroxyapatite nanowires. The highly flexible fire-resistant textiles possessed a similar morphology to the common cotton gauze (Figure 5.87A–D). In addition, the nonwoven highly flexible fire-resistant textile was also prepared using the nanoropes

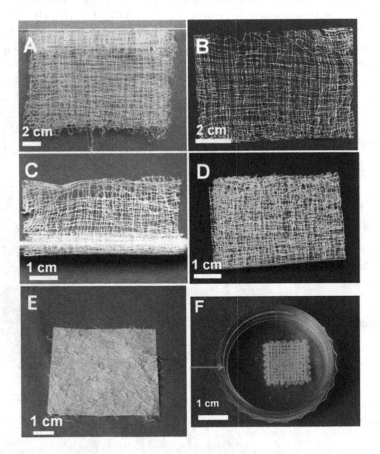

FIGURE 5.87 (A–D) Digital images of the as-prepared highly flexible fire-resistant textiles made from the nanoropes consisting of highly ordered ultralong hydroxyapatite nanowires; (E) a digital image of a nonwoven textile made from the nanoropes consisting of highly ordered ultralong hydroxyapatite nanowires; (F) a digital image of well-defined three-dimensional highly ordered pattern made from the nanoropes consisting of highly ordered ultralong hydroxyapatite nanowires using a commercial three-dimensional printer. (Reprinted with permission from reference [35])

consisting of highly ordered ultralong hydroxyapatite nanowires (Figure 5.87E). The well-defined highly ordered three-dimensional pattern constructed with the nano-ropes consisting of highly ordered ultralong hydroxyapatite nanowires was prepared using a commercial three-dimensional printer (Figure 5.87F). These highly flexible fire-resistant textiles may be engineered into advanced functional fire-resistant prod-ucts for applications in various fields, such as the fireproof clothing.

(3) Portable and Writable Photoluminescent Chalk for Information Protection

Information protection is particularly important for secure communication between parties, avoiding the leakage of secret information to unknown parties. Among vari-ous strategies used for information protection, security inks made from photolu-minescent materials are promising for various applications. However, the security inks have some disadvantages such as lack of universality on various substrates. The author's research group developed a new kind of portable and writable photo-luminescent chalk made from lanthanide-doped ultralong hydroxyapatite nanowires thermally treated at 1,000°C for 2 hours, which could directly write the covert infor-mation at anytime and anyplace on arbitrary surfaces and substrates, including paper sheets, metals, fabrics, plastics, woods, walls, foams, leaves, and even human body, as shown in Figure 5.88.[97] The written secret information was not visible under the ambient light, but it is readable upon exposure to ultraviolet light. Dual and triple encryption strategies were also developed for the high-level information protection.

(4) Fluorine-Free, Substrate-Independent Superhydrophobic Fire-Resistant Coatings

Environmentally harmful fluorinated compounds are frequently used as the main con-stituents for superhydrophobic coatings. The author's research group developed a simple, low-cost, and environmentally friendly production of fluorine-free, substrate-indepen-dent fire-resistant superhydrophobic coatings based on the superhydrophobic hydroxy-apatite nanowire paint.[98] The fire-resistant superhydrophobic paint could be coated on any substrate with any arbitrary shape by the spray-coating technique. The as-prepared coatings exhibited superior properties such as excellent fire resistance, high-temperature stability, superhydrophobicity, self-cleaning ability, and excellent mechanical durability. Furthermore, the fire-resistant superhydrophobic paint could be coated on large-sized practical objects to form large-area superhydrophobic fire-resistant coatings.

The fluorine-free, substrate-independent fire-resistant superhydrophobic coatings were also prepared on practical objects with large sizes and various shapes. Figure 5.89 shows the as-prepared fluorine-free, substrate-independent superhydrophobic fire-resistant coatings based on the superhydrophobic hydroxyapatite nanowire paint on various substrates with arbitrary shapes. Both flat (wood board and glass) and curved (plastic bottle) substrates were coated using the superhydrophobic hydroxyapatite nanowire paint by the spray-coating method. A flow of water was used to evaluate the water repellency of the large-area superhydrophobic fire-resistant coatings. The flow of water could rebound or roll off the coating surface, and no water stain remained on the

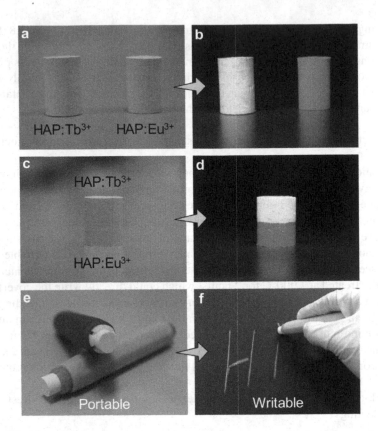

FIGURE 5.88 (a, b) The as-prepared portable and writable photoluminescent chalks doped with Eu³⁺ ions (red color under UV light irradiation) and Tb³⁺ ions (green color under UV light irradiation) under the ambient light with a white color (a) and 254 nm UV light (b); (c, d) the double-color photoluminescent chalk under the ambient light with a white color (c) and 254 nm UV light (top half was green, bottom half was red) (d); (e) the portable and writable photoluminescent chalks could be packed into the sleeves to carry conveniently; (f) the portable and writable photoluminescent chalk was writable on arbitrary substrates and surfaces. (Reprinted with permission from reference [97])

superhydrophobic coating. In addition, the mechanical stability of the large-area superhydrophobic fire-resistant coating was evaluated. A 70 kg man wearing shoes trod on the coated toughened glass. The experiments indicated that the superhydrophobic fire-resistant coating could be well preserved even after 500 steps.

In addition, a simple one-step calcium oleate precursor solvothermal method was developed to grow the ultralong hydroxyapatite nanowire coating on a glass substrate.[99] During the preparation process, the glass substrate was directly placed at the bottom of the Teflon-lined stainless-steel autoclave with the reaction suspension for solvothermal treatment at 180°C for 24 hours. The as-prepared hydroxyapatite nanowire coating showed excellent apatite-forming ability in simulated body fluid. It is expected that this method can also be extended to the preparation of ultralong hydroxyapatite nanowire

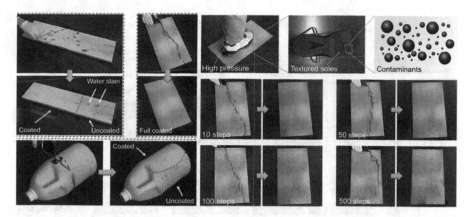

FIGURE 5.89 The as-prepared fluorine-free, substrate-independent superhydrophobic fire-resistant coatings based on the superhydrophobic hydroxyapatite nanowire paint on various substrates with arbitrary shapes. A 70 kg man wearing shoes trod on the toughened glass with a superhydrophobic fire-resistant coating for 500 steps. (Reprinted with permission from reference [98])

coatings on other kinds of substrate. The as-prepared ultralong hydroxyapatite nanowire coating is promising for various biomedical applications.

(5) Dental Enamel-Mimetic Large-Sized Multi-Scale Ordered Materials

Tooth enamel possesses excellent mechanical properties owing to its multi-scale, highly ordered architecture. To develop tooth enamel-mimetic structural materials is of great significance for understanding the important role of ordered hydroxyapatite nanocrystals in tooth enamel and exploring high-performance materials. However, it is still a great challenge to mimic the unique multi-scale ordered architecture of the natural tooth enamel. Inspired by the multi-scale structure and self-assembly of the tooth enamel, the author's research group developed a novel bottom-up step-by-step assembly strategy to build the tooth enamel-mimetic structural materials based on highly ordered ultralong hydroxyapatite nanowires reinforced with resin, and achieved the multi-scale (from nano- to micro- to macro-scale) highly ordered ultralong hydroxyapatite nanowire structure control, as shown in Figure 5.90.[100] Ultralong hydroxyapatite nanowires were assembled in sequence into highly ordered hydroxyapatite nanowire bundles, aligned hydroxyapatite microfibers, and three-dimensional highly ordered hydroxyapatite nanowire bulk, which could perfectly imitate the structure and highly ordered self-assembly of the tooth enamel. The unique multi-scale, highly ordered structure resulted in a significantly enhanced mechanical properties of the tooth enamel-mimetic structural materials based on highly ordered ultralong hydroxyapatite nanowires reinforced with resin, the compressive strength and compressive strain were 41.54 MPa and 1.71%, respectively; and the compressive Young's modulus was 2.80 GPa, which was 3.54 times that of the pure resin and 28 times that of the highly ordered hydroxyapatite nanowire

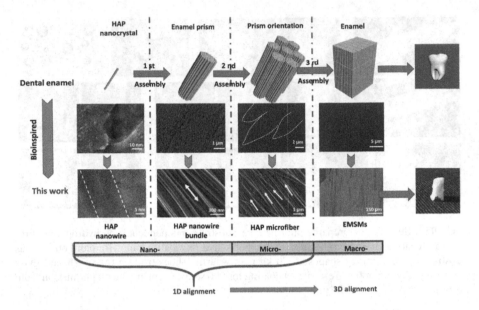

FIGURE 5.90 Schematic illustration of the structural similarity between the real tooth enamel and as-prepared biomimetic enamel. (Reprinted with permission from reference [100])

bulk sample. The greatly enhanced mechanical properties of the as-prepared tooth enamel-mimetic structural materials could result in the effective disruption step by step and energy consumption during the crack formation and propagation. This novel method is suitable to fabricate the tooth enamel-mimetic structural materials with arbitrary sizes and well-designed shapes, which are promising for various applications in the biomedical fields such as bone and tooth defect repair. By introducing other reinforcing materials and injection controlling programs, this strategy can be adopted for fabricating a variety of complex highly ordered hierarchical structured high-performance materials for various applications.

(6) Biocompatible, Highly Porous, Ultralight, Elastic Inorganic Aerogel

Inorganic aerogels have aroused great interest owing to their unique structures and superior properties. However, the practical applications of inorganic aerogels are greatly restricted by their high brittleness and high production cost. Inspired by the cancellous bone, the author's research group developed a new kind of inorganic aerogel based on ultralong hydroxyapatite nanowires with excellent elasticity, ultrahigh porosity (~99.7%), ultrasmall density (8.54 mg cm^{-3}, which is about 0.854% of water density), and low thermal conductivity (0.0387 W m^{-1} K^{-1}).[101] The as-prepared ultralong hydroxyapatite nanowires-based inorganic aerogel could be used as the highly efficient air filter with high filtration efficiencies for PM$_{2.5}$ and PM$_{10}$, and it could also be used as a smart switch to separate oil from water continuously. Compared with many other organic aerogels, the as-prepared ultralong

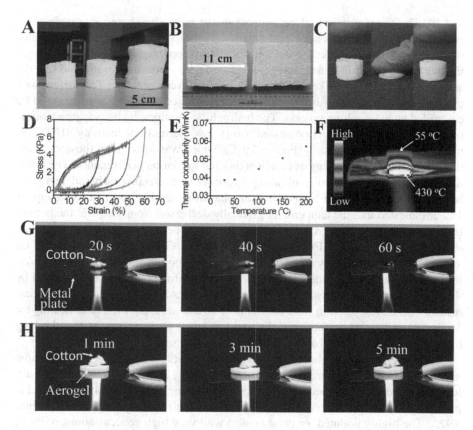

FIGURE 5.91 (A–C) Digital images of the hydrophobic ultralong hydroxyapatite nanowires-based inorganic aerogel samples with different sizes and shapes and excellent elasticity; (D) stress-strain curves of the hydrophobic ultralong hydroxyapatite nanowires-based inorganic aerogel during the compressing-releasing processes under different maximum strains (30%, 40%, 50%, and 60%); (E) thermal conductivities of the hydrophilic ultralong hydroxyapatite nanowires-based inorganic aerogel at different temperatures; (F) infrared image of the hydrophilic ultralong hydroxyapatite nanowires-based inorganic aerogel with a thickness of 1 cm on a metal plate after heating by an alcohol lamp for 1 hour; (G) a piece of cotton placed on a metal plate was heated by an alcohol lamp, and the cotton was burned within 1 minute; (H) a piece of cotton could be well protected by the hydrophilic ultralong hydroxyapatite nanowires-based inorganic aerogel (thickness ~5 mm) from the flame of an alcohol lamp even after heating for 5 minutes. (Reprinted with permission from reference [101])

hydroxyapatite nanowires-based inorganic aerogel is biocompatible, environmentally friendly, and low cost. In addition, the synthetic method can be scaled up for large-scale production of ultralong hydroxyapatite nanowires-based inorganic aerogel. Therefore, the ultralong hydroxyapatite nanowires-based inorganic aerogel is promising for applications in various fields.

As shown in Figure 5.91, the shape and size of the ultralong hydroxyapatite nanowires-based inorganic aerogel could be controlled using different molds (Figure 5.91A and B). The as-prepared ultralong hydroxyapatite nanowires-based inorganic aerogel exhibited an excellent elastic performance (Figure 5.91C). Figure 5.91D indicates that the maximum compressive strength of the hydrophobic ultralong hydroxyapatite nanowires-based inorganic aerogel increased with increasing strain, and that the aerogel exhibited a foam-like elasticity and shape recovery behavior. The surface wettability of the ultralong hydroxyapatite nanowires-based inorganic aerogel could be controlled by the washing process. The hydrophilic ultralong hydroxyapatite nanowires-based inorganic aerogel possessed a very low thermal conductivity (0.0387 W m^{-1} K^{-1} at room temperature) (Figure 5.91E), which was lower than those of many thermal insulators, indicating its excellent thermal insulation performance. The infrared image of the hydrophilic ultralong hydroxyapatite nanowires-based inorganic aerogel with a thickness of 1 cm on a metal plate after heating by an alcohol lamp for 1 hour indicated that the temperature gradually decreased from 430°C at the bottom to 55°C on the top of the aerogel, indicating excellent thermal insulation performance of the aerogel (Figure 6.91F). As shown in Figure 5.91G, a piece of cotton placed on a metal plate was heated by an alcohol lamp, and the cotton was burned and carbonized quickly within 1 minute. However, a piece of cotton could be well protected by the hydrophilic ultralong hydroxyapatite nanowires-based inorganic aerogel as a heat insulator (thickness ~5 mm) from the flame of an alcohol lamp even after heating for 5 minutes, indicating the superior thermal insulation performance of the ultralong hydroxyapatite nanowires-based inorganic aerogel (Figure 5.91H).

The as-prepared hydrophobic ultralong hydroxyapatite nanowires-based inorganic aerogel possessed high filtration efficiencies for $PM_{2.5}$ and PM_{10} in the polluted air, low pressure drop, good mechanical properties, and high biocompatibility; thus, it is promising for the application in breathing masks and air purifiers, as shown in Figure 5.92A. The highly polluted air (white color) with very high concentrations of $PM_{2.5}$ and PM_{10} was generated in a hermetic chamber, and the white color turned to almost clear after the air purification process for 12 minutes by the hydrophobic ultralong hydroxyapatite nanowires-based inorganic aerogel filter, indicating its excellent performance in the air purification (Figure 5.92B), and at the same time the aerogel filter obviously changed from white to brown yellow after the air purification process for 2 hours (Figure 5.92C), indicating that a large number of $PM_{2.5}$ and PM_{10} particles were captured by the aerogel filter. The captured particulate matter particles on the aerogel filter could be collected by dissolving the aerogel filter in the hydrochloric acid aqueous solution (Figure 5.92D and E). These experimental results demonstrated that the as-prepared hydrophobic ultralong hydroxyapatite nanowires-based inorganic aerogel is promising for applications in air purifiers, breathing masks, etc.[101]

The hydrophobic ultralong hydroxyapatite nanowires-based inorganic aerogel possessed a high hydrophobicity with an average water contact angle of 144.8°, which was desirable for the highly efficient absorption of oils and organic solvents. Figure 5.93A shows high absorption capacities of the hydrophobic ultralong hydroxyapatite nanowires-based inorganic aerogel for different oils and organic solvents, and the amounts of absorbed oils or organic solvents were 83–156 times the weight of the aerogel. In addition, the excellent elasticity of the hydrophobic ultralong hydroxyapatite nanowires-based inorganic aerogel could enable the excellent recyclability and

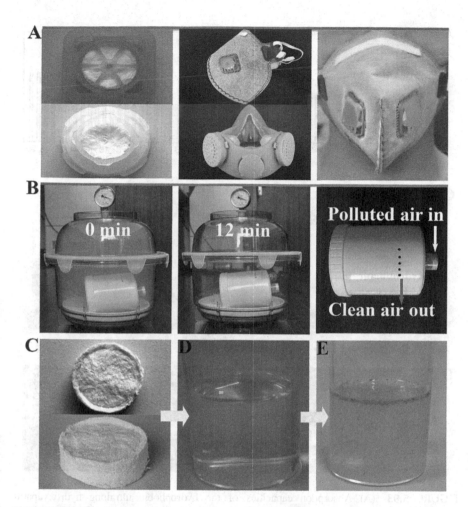

FIGURE 5.92 (A) Different kinds of breathing masks prepared using the hydrophobic ultralong hydroxyapatite nanowires-based inorganic aerogel filter; (B) digital images of a hermetic chamber filled with highly polluted air with a very high concentration of $PM_{2.5}$, and a prototype filter model prepared using the hydrophobic ultralong hydroxyapatite nanowires-based inorganic aerogel for the air purification for 12 minutes; (C) digital images of the hydrophobic ultralong hydroxyapatite nanowires-based inorganic aerogel filter used in the prototype filter model after the air purification process for 2 hours; (D) digital image of the hydrochloric acid aqueous solution; (E) digital image of the hydrochloric acid aqueous solution after dissolution of the hydrophobic ultralong hydroxyapatite nanowires-based inorganic aerogel filter used for air purification for 2 hours. (Reprinted with permission from reference [101])

stable oil/solvent absorption capacity, which was demonstrated by the absorption-squeeze-absorption recycling tests (Figure 5.93B). After 20 cycles, the chloroform absorption capacity was still 120 times the weight of the aerogel, revealing the foam-like behavior and excellent recycling performance of the hydrophobic ultralong hydroxyapatite nanowires-based inorganic aerogel.

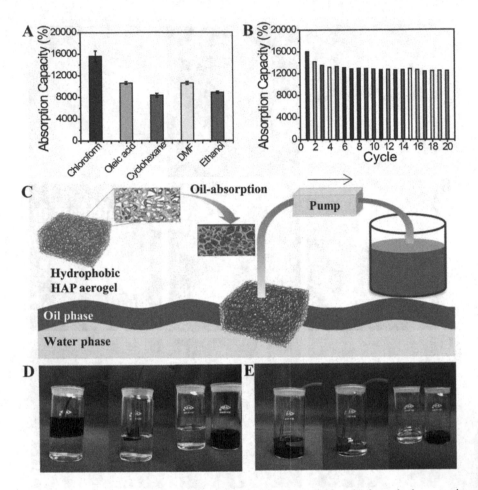

FIGURE 5.93 (A) Absorption capacities of the hydrophobic ultralong hydroxyapatite nanowires-based inorganic aerogel for different oils and organic solvents (DMF stands for N,N-dimethylformamide); (B) absorption capacities of the hydrophobic ultralong hydroxyapatite nanowires-based inorganic aerogel for chloroform for different cycles; the absorbed oil or organic solvent in the aerogel was squeezed out by hand, and the aerogel was reused after each cycle; (C) schematic illustration of the continuous oil–water separation device prepared using the hydrophobic ultralong hydroxyapatite nanowires-based inorganic aerogel; (D) continuous separation of low-density oil (cyclohexane dyed with oil red) by the as-prepared oil–water separation device; (E) continuous separation of high-density oil (chloroform dyed with oil red) by the as-prepared oil–water separation device. (Reprinted with permission from reference [101])

A prototype separation device for continuous oil–water separation was designed. The as-prepared oil–water separation device consisted of a piece of hydrophobic ultralong hydroxyapatite nanowires-based inorganic aerogel, a tube, and a peristaltic pump, and the tube was inserted into the aerogel and immersed into the oil–water mixture. This oil–water separation device could rapidly separate oil from water conveniently and continuously (Figure 5.93C–E). Both low-density oil (cyclohexane, Figure 5.93D) and high-density oil (chloroform, Figure 5.93E) could be rapidly and continuously separated from the water–oil mixture and collected in a container by the oil–water separation device. In this device, the hydrophobic ultralong hydroxyapatite nanowires-based inorganic aerogel acted as a smart reservoir to absorb oil and allowed only oil to pass through the aerogel. This novel method is fast, continuous, low cost, and environmentally friendly, in contrast to many other materials which use heat treatment to remove absorbed oils.[101]

As discussed above, hydroxyapatite is promising for applications in bone defect repair and bone regeneration. However, highly porous hydroxyapatite scaffolds for bone defect repair usually exhibit high brittleness and poor mechanical properties, thus organic constituents are usually added to form composite materials. The highly porous elastic ultralong hydroxyapatite nanowires-based inorganic aerogel with an ultrahigh porosity, excellent elasticity, and suitable porous structure could be used as the high-performance scaffold for bone defect repair. The highly porous structure of the as-prepared highly porous elastic ultralong hydroxyapatite nanowires-based inorganic aerogel was desirable for bone ingrowth and matter/fluid transfer, and the high elasticity and good mechanical properties could ensure its well fitting into the irregularly shaped bone defect cavity and structural integrity of the scaffold during bone regeneration process. The highly porous elastic ultralong hydroxyapatite nanowires-based inorganic aerogel scaffold could promote the adhesion, proliferation, and migration of rat bone marrow–derived mesenchymal stem cells, and enhance the protein expression of osteogenesis and angiogenesis related genes. The *in vivo* experimental results indicated that the highly porous elastic ultralong hydroxyapatite nanowires-based inorganic aerogel scaffold exhibited high performance for the ingrowth of new bone and blood vessels, and significantly accelerate new bone growth and neovascularization.[102]

As shown in Figure 5.94, a critical-sized rat calvarial defect model was used to evaluate the efficacy of the highly porous elastic ultralong hydroxyapatite nanowires-based inorganic aerogel scaffold for *in vivo* bone defect repair. After three months post-implantation, the obvious new bone formation was observed in the bone defect region for the highly porous elastic inorganic aerogel scaffold group, and the newly formed bone grew deep into the aerogel scaffold without any obvious crack. In contrast, less new bone was observed on the surface of the hydroxyapatite nanorod ceramic scaffold, and almost no new bone formation was observed in the bone defect region without any scaffold (Figure 5.94A). The ratio of bone volume to total volume (BV/TV) and bone surface density (BS/TV) were obviously higher in the highly porous and elastic inorganic aerogel scaffold group ($31.83 \pm 7.21\%$ and 2.96 ± 0.33, respectively), compared with the hydroxyapatite nanorod ceramic group ($15.42 \pm 4.09\%$ and 0.99 ± 0.23, respectively), and the blank control group ($6.87 \pm 1.61\%$ and 0.77 ± 0.24, respectively). Although the difference of BS/TV between the hydroxyapatite nanorod ceramic group and the blank control group was not significant, the BV/TV of the hydroxyapatite nanorod ceramic group was approximately two times that of the control group (Figure 5.94C and D).

In addition to bone regeneration, an ideal porous scaffold should have high ability to promote the angiogenesis. More new blood vessels were observed in the bone

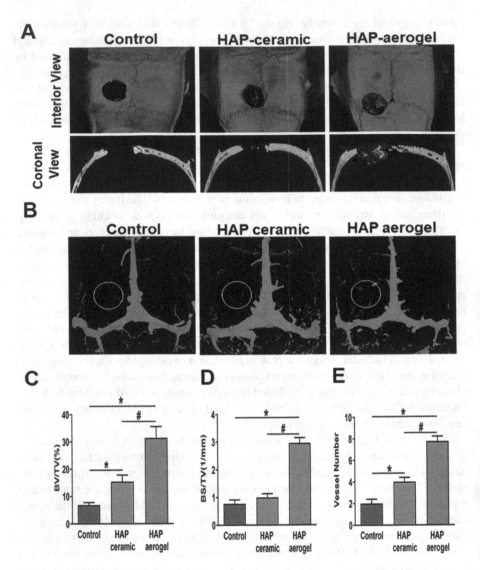

FIGURE 5.94 Micro-CT assessment of newly formed bone and blood vessels in rat calvarial defect regions after implantation for 12 weeks of the highly porous elastic ultralong hydroxyapatite nanowires-based inorganic aerogel scaffold, and the hydroxyapatite nanorod ceramic scaffold prepared by sintering at 1,200°C for 2 hours. (A) Three-dimensional and coronal views of reconstructed calvarial images; (B) three-dimensional reconstructed images of newly formed blood vessels after Microfil perfusion; (C–E) bone volume (BV)/total volume (TV) (C), bone surface density (BS/TV) (D), and blood vessel number (E) in the bone defect areas. (Reprinted with permission from reference [102])

defect region with the highly porous elastic ultralong hydroxyapatite nanowires-based inorganic aerogel scaffold compared with the hydroxyapatite nanorod ceramic scaffold and the control group (Figure 5.94B), indicating that the highly porous elastic ultralong hydroxyapatite nanowires-based inorganic aerogel scaffold could promote the angiogenesis in the bone defect region. The blood vessel number of the highly porous elastic ultralong hydroxyapatite nanowires-based inorganic aerogel scaffold (7.75 ± 0.96) was higher than those of the hydroxyapatite nanorod ceramic scaffold (4.00 ± 0.81) and the control group (2.00 ± 0.82) (Figure 5.94E).[102]

(7) Elastic Porous Nanocomposite Scaffolds for Bone Defect Repair

Millions of people worldwide need bone grafts to reconstruct large bone defects caused by trauma, tumor surgery, and infection each year. Nowadays, autologous bone grafting and allogeneic bone grafting are the desirable treatments for large bone defects. However, the clinical applications of the autograft and allograft are limited by the donor-site morbidity, limited supply, and risks of infection and immunological reaction. To solve these problems, various types of synthetic bone graft substitutes were studied for bone regeneration *in vivo*. To achieve successful bone regeneration, the synthetic porous scaffolds for bone defect repair should be osteoconductive and osteoinductive; that is, the ideal bone repair scaffolds should facilitate bone-related cell proliferation and osteogenic differentiation, support the formation of new bone tissue and blood vessels, and be gradually metabolized and replaced by the newly formed tissues. In addition, the ideal synthetic bone graft substitute should possess excellent mechanical properties to allow surgical handling and fixation.

The synthetic hydroxyapatite is similar to the major inorganic component of the natural bone in terms of chemical composition and structure, and has high biocompatibility, excellent cell adhesion, and high osteoconductivity. Moreover, collagen, as the organic constituent of the bone extracellular matrix, is a promising biomaterial owing to its excellent biocompatibility, cell affinity, and biodegradability. As a result, the hydroxyapatite/collagen composite scaffolds can chemically and structurally mimic the natural bone, and have excellent bioactivity, biodegradability, and osteoinductive activity. In addition, the porous structure of the composite scaffolds with high porosity is beneficial for cell infiltration, new bone ingrowth, and neovascularization.

Inspired by the chemical composition and structure of the natural bone, a biomimetic elastic porous nanocomposite scaffold consisting of 66.7 wt.% ultralong hydroxyapatite nanowires and collagen with a high porosity (96.61%) was synthesized by the freeze drying and subsequent chemical crosslinking.[103, 104] Compared with the pure collagen control sample, the elastic porous nanocomposite scaffold consisting of 66.7 wt.% ultralong hydroxyapatite nanowires and collagen showed greatly enhanced mechanical properties. Furthermore, the rehydrated elastic porous nanocomposite scaffold possessed excellent elastic performance, as shown in Figure 5.95a–d. The elastic porous nanocomposite scaffold exhibited a good degradable performance with a sustainable release of Ca and P elements, and could promote the adhesion and spreading of mesenchymal stem cells. The *in vivo* evaluation indicated that the

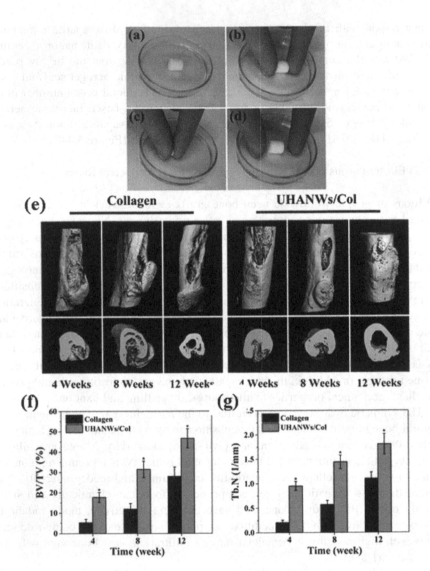

FIGURE 5.95 (a–d) Digital images showing the excellent elastic performance of the as-prepared biomimetic elastic porous nanocomposite scaffold consisting of 66.7 wt.% ultralong hydroxyapatite nanowires and collagen rehydrated in deionized water; (e–g) Micro-CT analysis results of newly formed bone in the rabbit radialis defect region implanted with the collagen sample and the biomimetic elastic porous nanocomposite scaffold for 4-week, 8-week, and 12-week post-implantation: (e) three-dimensional reconstructed rabbit radialis images; (f) the percentage of new bone volume to tissue volume (BV/TV); (g) trabecular number (Tb.N). (Reprinted with permission from reference [103])

elastic porous nanocomposite scaffold could significantly enhance bone regeneration compared with the pure collagen sample. After 12 weeks of implantation, the new bone was formed throughout the entire bone defect area, and the newly formed bone could connect directly with the host bone to construct a relatively normal bone marrow cavity, leading to successful osteointegration and bone regeneration. As shown in Figure 5.95e, more new bone formation was observed in the radialis defect region of the collagen sample group and biomimetic elastic porous nanocomposite scaffold group with increasing implantation time. Compared with the collagen control group, much more new bone formed in the biomimetic elastic porous nanocomposite scaffold group. Furthermore, a relatively normal bone marrow cavity was observed in the biomimetic elastic porous nanocomposite scaffold group after 12 weeks post-implantation. In contrast, the bone defect repairing performance of the collagen control group was not satisfactory, resulting in the formation of an irregular bone marrow cavity at 12 weeks post-implantation.

The quantitative analysis of the rabbit radialis defect region indicated that the percentage of new bone volume to tissue volume (BV/TV) was $(5.24 \pm 1.73)\%$, $(11.86 \pm 2.81)\%$, and $(27.95 \pm 4.59)\%$, respectively, for the collagen control group after 4, 8, and 12 weeks post-implantation (Figure 5.95f), the corresponding trabecular number (Tb.N) was 0.20 ± 0.05 mm^{-1}, 0.57 ± 0.08 mm^{-1}, and 1.10 ± 0.13 mm^{-1}, respectively (Figure 5.95g). In comparison, the BV/TV in the biomimetic elastic porous nanocomposite scaffold group was $(17.67 \pm 2.45)\%$, $(31.46 \pm 4.01)\%$, and $(46.61 \pm 4.74)\%$, respectively (Figure 5.95f), and the corresponding Tb.N in the biomimetic elastic porous nanocomposite scaffold group was 0.95 ± 0.09 mm^{-1}, 1.44 ± 0.14 mm^{-1}, and 1.81 ± 0.21 mm^{-1}, respectively, after 4 weeks, 8 weeks, and 12 weeks post-implantation (Figure 5.95g), which were obviously higher than those in the collagen control group.[103]

In addition, the metal ion–doped or hybridized ultralong hydroxyapatite nanowires and their derived porous scaffolds are also promising for the application in bone defect repair and bone regeneration.[105, 106] Furthermore, based on ultralong hydroxyapatite nanowires, hydroxyapatite nanowire/magnesium silicate nanosheet core-shell porous hierarchical scaffold with high-performance drug loading and sustained release was prepared for bone defect repair.[107]. The as-prepared core-shell porous hierarchical scaffold was further integrated into the chitosan matrix. The core-shell porous hierarchical scaffold could promote the attachment and growth of rat bone marrow–derived mesenchymal stem cells, and induce the expression of osteogenic differentiation related genes and the vascular endothelial growth factor genes. Moreover, the core-shell porous hierarchical scaffold could significantly stimulate *in vivo* bone regeneration and *in vivo* angiogenesis.[107]

(8) Highly Porous Ceramics

In recent years, hydroxyapatite porous ceramics as the promising candidate for bone defect repair have become a hot research topic owing to their high biocompatibility and osteoconductivity. The bone defect repair performance of the biomedical porous scaffolds depends on their chemical composition, structure, pore size, porosity, mechanical properties, etc. In order to promote the bone regeneration, the scaffolds for bone defect repair should have the porous structure to facilitate the nutrient transportation, cell infiltration, new bone ingrowth, and neovascularization. Generally speaking, materials made from one-dimensional building blocks exhibit superior mechanical properties and porous structure. Most of hydroxyapatite porous ceramics reported in the literature were prepared using hydroxyapatite particles, and only a small number of porous ceramics were prepared using hydroxyapatite fibers.

The author's research group prepared the highly porous ceramics using ultralong hydroxyapatite nanowires as the building blocks and palmitic acid spheres as the pore former.[108] During the sintering process, although palmitic acid spheres were decomposed at high temperatures, ultralong hydroxyapatite nanowires could be well preserved and still support the whole porous structure stably, resulting in the formation of the highly porous ceramics with good mechanical properties. The as-prepared highly porous ceramics possessed high biocompatibility, superior performance for cell adhesion, spreading and cell proliferation. Thus, the as-prepared highly porous

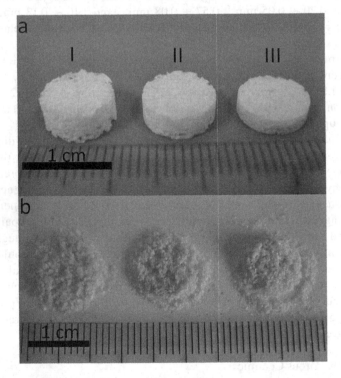

FIGURE 5.96 Digital images of the ultralong hydroxyapatite nanowires-based highly porous ceramics (a) and the products of hydroxyapatite nanoparticles after sintering (b) prepared by using different amounts of palmitic acid spheres: (I) 350 mg; (II) 250 mg; (III) 150 mg and sintering at 1300°C for 4 hours. (Reprinted with permission from reference [108])

ceramics based on ultralong hydroxyapatite nanowires are promising for applications in various biomedical fields.

Figure 5.96 shows digital images of the ultralong hydroxyapatite nanowires-based highly porous ceramics in comparison with the products of hydroxyapatite nanoparticles prepared by using different amounts of palmitic acid spheres as the pore former. The highly porous hydroxyapatite nanowire porous ceramics could be successfully prepared in all three cases, and the porosity and volume of the hydroxyapatite nanowire porous ceramic increased with increasing amount of palmitic acid spheres (Figure 5.96a). In contrast, the green bodies of hydroxyapatite nanoparticles collapsed to powders during the sintering process in all three cases (Figure 5.96b). The experimental results indicated the important role of ultralong hydroxyapatite nanowires in the formation of high-performance highly porous ceramics.

(9) Hydrogels

In recent years, hydrogels have attracted much attention owing to their applications in repairing and regenerating tissues and organs. Hydrogels are composed of the water-soluble polymer network, and the hydrophilic part combines with water molecules and keeps water inside the network. The polymer materials can be naturally occurring, synthetic or hybrid to satisfy specific mechanical and chemical properties. Hydrogels with three-dimensional cross-linked structures are widely used in various biomedical fields such as bone defect repair scaffolds, drug carriers and biosensors. Hydrogels can influence the adhesion, geometry, and proliferation of the cells when they serve as the mimics for the extracellular matrix. The migration, development, and differentiation of the cells are sensitive to the elastic modulus of hydrogels. As a naturally occurring material, sodium alginate is an important biomaterial with high biocompatibility. However, the mechanical properties of sodium alginate-based hydrogels are usually poor, which restricts their applications. To solve this problem, the hybridization of organic and inorganic constituents to prepare composite hydrogels is a useful strategy.

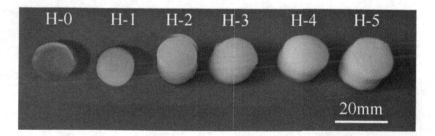

FIGURE 5.97 Digital images of the as-prepared ultralong hydroxyapatite nanowire/sodium alginate porous hydrogel samples with different weight ratios of ultralong hydroxyapatite nanowire/sodium alginate (from left to right: 100% sodium alginate, 1:5, 1:3, 1:2, 1:1, 2:1). (Reprinted with permission from reference [109])

The author's research group developed ultralong hydroxyapatite nanowire/sodium alginate porous hydrogels with superior mechanical properties using divalent calcium ions as the crosslinking agent, as shown in Figure 5.97.[109] The as-prepared porous hydrogels had a porous structure with pore sizes ranging from ~200 μm to ~500 μm. The mechanical properties of porous hydrogels could be greatly enhanced by the addition of ultralong hydroxyapatite nanowires. The maximum compressive modulus and tensile Young's modulus of the as-prepared porous hydrogel (ultralong hydroxyapatite nanowire/ sodium alginate weight ratio = 2:1) were 0.123 MPa and 0.994 MPa, which were ~1.62 and 6.14 times those of the pure sodium alginate hydrogel, respectively. Owing to enhanced mechanical properties and high biocompatibility, the as-prepared porous hydrogels have promising applications in various biomedical fields, such as bone defect repair.

6 Acid/Alkali-Proof Fire-Resistant Paper Based on Barium Sulfate Fibers

The environment friendly fire-resistant inorganic paper made from ultralong hydroxyapatite nanowires has many advantages, such as high biocompatibility, good flexibility, being writable and printable, and excellent resistance to harsh conditions such as fire, high temperatures, and strong alkalis. However, although the ultralong hydroxyapatite nanowire-based fire-resistant paper has excellent resistance to strong alkalis, it cannot resist strong acids, and this weakness limits its applications in strong acid environments. Up until now, it remains a great challenge to prepare the high-performance fire-resistant inorganic paper with combined merits of high biocompatibility, high flexibility, excellent mechanical properties, and excellent resistance to fire, high temperatures, strong acids, and strong alkalis.

Barium sulfate is a promising building material for the acid/alkali-proof fire-resistant paper because it has a high melting point of 1580°C and high chemical stability in strong acids and strong alkalis. In addition, barium sulfate has good biosecurity, and it is widely used as the gastrointestinal contrast agent in medical tests. The author's research group developed a new kind of acid/alkali-proof fire-resistant inorganic paper comprising fibers self-assembled with barium sulfate nanorods prepared by the barium oleate precursor hydrothermal method in a large scale.[110] The as-prepared barium sulfate fibers have a three-level hierarchical structure: (1) barium sulfate nanorods; (2) fine fibers self-assembled with barium sulfate nanorods; (3) thick fibers self-assembled with fine barium sulfate fibers, as shown in Figure 6.1. The as-prepared acid/alkali-proof fire-resistant inorganic paper has high flexibility, good mechanical strength, and excellent resistance to harsh conditions such as fire, high temperatures, strong acids, and strong alkalis. The as-prepared barium sulfate fibers had lengths up to several hundred micrometers and diameters of ~1 μm, and were composed of parallel self-assembled nanorods with diameters of ~20 nm and lengths of 50–150 nm. It was a directional alignment of oleate groups on the surface of barium sulfate nanorods with the hydrophobic end toward outside, resulting in the self-assembly of barium sulfate nanorods along the longitudinal direction to form barium sulfate fibers. The self-assembly of barium sulfate nanorods occurred during the hydrothermal process, and this could be proved by the fact that barium sulfate fibers with lengths of several hundred of micrometers were formed in the hydrothermal product slurry without washing.

The digital images of the as-prepared acid/alkali-proof fire-resistant inorganic paper sheets are shown in Figure 6.2A and B. Sample 1 was a pure barium sulfate

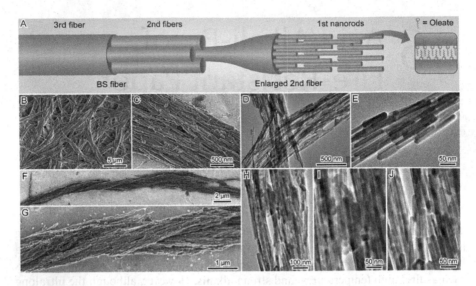

FIGURE 6.1 Characterization of the three-level hierarchical structure of the as-prepared fibers formed by self-assembly of barium sulfate nanorods. (A) Schematic illustration of the three-level hierarchical structure of a fiber formed by self-assembly of barium sulfate nanorods; (B, C, F, G) SEM images; (D, E, H, I, J) TEM images. (Reprinted with permission from reference [110])

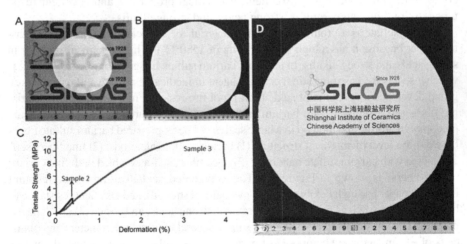

FIGURE 6.2 Characterization of the acid/alkali-proof fire-resistant inorganic paper based on fibers self-assembled with barium sulfate nanorods. (A) A digital image of the pure barium sulfate fiber paper (sample 1) with a diameter of 4 cm (paper thickness ~100 μm); (B) digital image of the acid/alkali-proof fire-resistant inorganic paper (paper thickness ~350 μm) with a diameter of 20 cm (sample 2) and 4 cm (sample 3); (C) tensile stress–strain curves of the acid/alkali-proof fire-resistant inorganic paper sheets with a diameter of 20 cm (sample 2) and 4 cm (sample 3); (D) a digital image of the acid/alkali-proof fire-resistant inorganic paper (paper thickness ~190 μm, diameter 20 cm) after color printing using a commercial ink-jet printer. (Reprinted with permission from reference [110])

fiber paper consisting of only barium sulfate fibers, samples 2 and 3 were prepared using different weight ratios of barium sulfate fibers, glass fibers, and inorganic adhesive. Sample 2 had a tensile strength of 2.5 MPa, and sample 3 exhibited a much higher tensile strength of 11.9 MPa and a deformation percentage of about 4% (Figure 6.2C). Therefore, the mechanical properties of the acid/alkali-proof fire-resistant inorganic paper could be adjusted by varying the composition of the paper. The as-prepared acid/alkali-proof fire-resistant inorganic paper was white, flexible, and smooth on the surface, and it could be used for writing and color printing. As shown in Figure 6.2D, the text and graphic pattern with various colors could be finely printed on the acid/alkali-proof fire-resistant inorganic paper.

The fire-resistant performance of the acid/alkali-proof fire-resistant inorganic paper was investigated, as shown in Figure 6.3. The acid/alkali-proof fire-resistant inorganic paper caught fire in 0.5 second once it was placed on the flame of a spirit lamp due to adsorbed oleate groups on the surface of barium sulfate fibers. The fire could extinguish rapidly after adsorbed oleate groups were burned out. The color of the acid/alkali-proof fire-resistant inorganic paper turned from white to gray because of the formation of carbon particles (Figure 6.3A–D). Furthermore, in order to test the high-temperature flexibility of the acid/alkali-proof fire-resistant inorganic paper, the paper underwent bending–stretching cycles on the flame of a spirit lamp (Figure 6.3E–P).

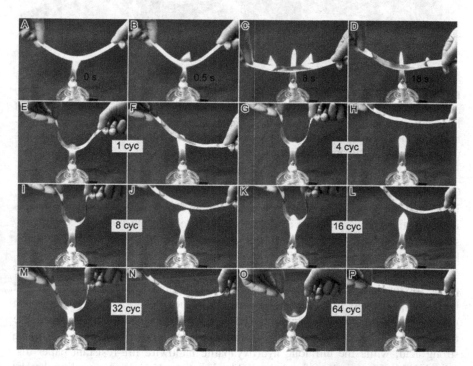

FIGURE 6.3 Fire-resistant and flexible properties of the acid/alkali-proof fire-resistant inorganic paper. (A–D) Fire-resistant tests; (E–P) cycled bending–stretching tests on the flame of a spirit lamp. (Reprinted with permission from reference [110])

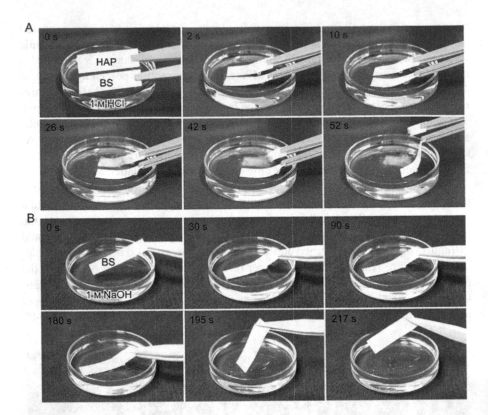

FIGURE 6.4 Corrosion resistance tests to a strong acid and a strong alkali of the acid/alkali-proof fire-resistant inorganic paper. (A) The resistance test to a strong acid in 1 M hydrochloric acid, in comparison with a fire-resistant paper made from ultralong hydroxyapatite nanowires; (B) the resistance test to a strong alkali in 1 M NaOH aqueous solution. (Reprinted with permission from reference [110])

The acid/alkali-proof fire-resistant inorganic paper could well maintain its high flexibility and integrity after 64 bending–stretching cycles. These experimental results indicated that the as-prepared acid/alkali-proof fire-resistant inorganic paper possessed excellent fire-resistant performance as well as high flexibility even under high-temperature conditions.

The acid-resistant performance of the acid/alkali-proof fire-resistant inorganic paper was evaluated, as shown in Figure 6.4. The acid/alkali-proof fire-resistant inorganic paper was immersed in 1 M hydrochloric acid solution, in comparison with a fire-resistant paper consisting of ultralong hydroxyapatite nanowires (Figure 6.4A). The acid/alkali-proof fire-resistant inorganic paper could preserve well in the strong acid, while the ultralong hydroxyapatite nanowire fire-resistant paper dissolved within 52 seconds, and only insoluble glass fibers remained, revealing that the acid/alkali-proof fire-resistant inorganic paper exhibited excellent corrosion resistance to the strong acid. In addition, the alkali-resistant performance of the acid/

alkali-proof fire-resistant inorganic paper was also tested. The acid/alkali-proof fire-resistant inorganic paper was immersed in 1 M sodium hydroxide aqueous solution, and the paper could be well preserved after 217 seconds, indicating that the acid/alkali-proof fire-resistant inorganic paper possessed excellent corrosion resistance to the strong alkali (Figure 6.4B).

7 Commercialization Potential and Future Prospects of the Fire-Resistant Paper

As discussed above, the fire-resistant paper based on ultralong hydroxyapatite nanowires has many advantages which traditional plant fiber paper does not possess; for example, the fire-resistant paper based on ultralong hydroxyapatite nanowires is environment friendly, has a high-quality natural white color without bleaching and sizing, and the most amazing thing is that it is highly resistant to both fire and high temperatures, and is expected to be used for long-term safe preservation of important documents such as archives and books. In addition, the fire-resistant paper also has many other uses and has good application prospects in various fields. The ultralong hydroxyapatite nanowires, which are the raw material of the new type of fire-resistant paper, can be synthesized artificially using common chemicals without consuming valuable natural resources such as trees. The whole manufacturing process of the new fire-resistant paper is environment friendly and will not pollute the environment. Thus, the new fire-resistant paper has a tempting prospect for commercialization and large-scale applications. It can be expected that in the near future the new fire-resistant paper may be able to move from the laboratory to the market, into book stores and libraries, and to protect important documents, archives, and books, and for applications in many other fields.

The commercialization potential and future prospects of the new kind of fire-resistant paper based on ultralong hydroxyapatite nanowires mainly depend on the development of large-scale, low-cost, environment-friendly synthetic methods for the industrial production of the building material of ultralong hydroxyapatite nanowires. In 2013, the author's research group developed the calcium oleate precursor solvothermal method for the synthesis of ultralong hydroxyapatite nanowires with diameters of about 10 nm and lengths of several hundred micrometers and aspect ratios of >10,000 using water-soluble calcium salt, phosphate salt, and alkali in mixed solvents of water, alcohol, and oleic acid by the solvothermal method.[12, 31, 32] In 2017, the author's research group developed a more environmentally friendly and low-cost calcium oleate precursor hydrothermal method for the synthesis of network-structured ultralong hydroxyapatite nanowires using water-soluble calcium salt, sodium oleate, and water-soluble phosphate in water as the only solvent without using any organic solvent.[34] Other compounds with similar structures and properties to sodium oleate, such as sodium stearate and sodium laurate, could also be used to synthesize network-structured ultralong hydroxyapatite nanowires by the calcium oleate precursor hydrothermal method.

The large-scale production of nanostructured materials is a universal challenge and a bottleneck for practical applications of nanostructured materials. The laboratory synthesis of nanostructured materials is usually in a small scale (less than 100 mL reaction systems). Currently, the author's research group has developed the scaled-up calcium oleate precursor solvothermal/hydrothermal technology for the synthesis of ultralong hydroxyapatite nanowires using stainless-steel autoclaves with volumes of 10 L and 100 L in the laboratory.

The production of ultralong hydroxyapatite nanowires needs to be further scaled up to the industrial production scale in the future. Before the realization of the industrial production, pilot tests need to be carried out to evaluate and optimize the synthetic technologies of ultralong hydroxyapatite nanowires. In addition, new environment-friendly, low-cost, and large-scale synthetic technologies should be explored to meet the high requirements of the industrial production of ultralong hydroxyapatite nanowires.

On the other hand, technologies for the rapid and continuous production of the fire-resistant paper based on ultralong hydroxyapatite nanowires need to be developed. Papermaking machines are needed for the rapid and continuous production of the fire-resistant paper based on ultralong hydroxyapatite nanowires. However, traditional cellulose fiber paper–making machines are not suitable for the rapid and continuous production of the fire-resistant paper based on ultralong hydroxyapatite nanowires because of the significant differences between cellulose fibers and ultralong hydroxyapatite nanowires. Therefore, new types of papermaking machines should be designed and developed for the rapid and continuous production of the fire-resistant paper based on ultralong hydroxyapatite nanowires.

Another main research direction currently being performed in the author's research laboratory is the low-cost, low-temperature open reaction system synthesis of calcium phosphate fibers as the building material of the fire-resistant inorganic paper. As discussed above, the previously developed calcium oleate precursor solvothermal/hydrothermal method for the synthesis of ultralong hydroxyapatite nanowires is carried out in the closed reaction systems under high temperatures and high pressures. The apparatus used in the calcium oleate precursor solvothermal/hydrothermal synthesis is stainless-steel autoclave, which is relatively expensive and not easy to be scaled up to a large industrial scale. The low-cost, low-temperature open reaction system synthesis of calcium phosphate fibers can be carried out in the aqueous solution in the open vessels at low temperatures below 100°C, and less electricity is consumed, and the open reaction vessels are much cheaper than the closed reaction system stainless-steel autoclaves. In addition, the author's research group uses cheap tap water instead of expensive deionized water, and cheap industrial chemical raw materials instead of expensive analytically pure chemicals for the synthesis of calcium phosphate fibers in the open reaction systems. Therefore, the production cost of calcium phosphate fibers in the open reaction systems can be significantly lowered. A typical example of calcium phosphate fibers synthesized in the open reaction

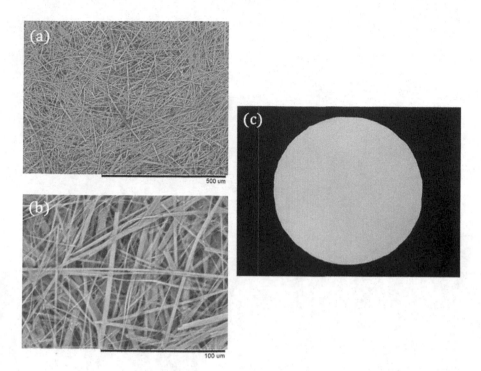

FIGURE 7.1 (a, b) SEM images of calcium phosphate fibers synthesized by a low-cost, low-temperature open reaction system synthetic method; (c) digital image of the fire-resistant paper with a diameter of 20 cm prepared using calcium phosphate fibers as the building material synthesized by the low-cost, low-temperature open reaction system synthetic method.

system is shown in Figure 7.1a and b; the calcium phosphate fibers possessed diameters of several micrometers and lengths of several hundred micrometers. The calcium phosphate fibers synthesized in the open reaction system could be used as building material for the preparation of the fire-resistant paper (Figure 7.1c). The fire-resistant paper based on calcium phosphate fibers exhibited similar excellent fire and high-temperature resistance performance as the fire-resistant paper based on ultralong hydroxyapatite nanowires. Therefore, the environmentally friendly, low-cost, low-temperature, open reaction system synthetic method for calcium phosphate fibers is very promising for the industrial scale production and large-scale applications of the fire-resistant inorganic paper. It is expected that with sufficiently low production cost and mature large-scale industrial synthetic technology, the new kind of fire-resistant inorganic paper will become a commodity for various applications in the market and enter into the daily work and life of ordinary people in the future.

References

1. Papyrus. *New World Encyclopedia* 2019, https://www.newworldencyclopedia.org/p/index.php?title=Papyrus&oldid=1017184.
2. History of paper. *Wikipedia, the free encyclopedia* 2021, https://en.wikipedia.org/wiki/History_of_paper.
3. M. A. Hubbe, C. Bowden. Handmade paper: a review of its history, craft, and science. *BioResources* 2009, *4*, 1736–1792.
4. D. Hunter. *Papermaking: the history and technique of an ancient craft.* Alfred A. Knopf, New York 1947.
5. E. Koretsky. *Killing green: an account of hand papermaking in China.* Legacy Press, Ann Arbor, Michigan 2009.
6. M. H. Hart. *The 100: a ranking of the most influential persons in history.* Citadel Press, Kensington Publishing Corp., New York 1992.
7. Best inventors. *Time* 2007, November 2 issue.
8. T. H. Barrett. *The woman who discovered printing.* Yale University Press 2008.
9. T.-H. Tsien. Paper and printing. Joseph Needham ed., *Science and Civilisation in China, Chemistry and Chemical Technology,* Cambridge University Press 1985, Volume 5.
10. E. Wilkinson. *Chinese history: a new manual.* Harvard University Asia Center for the Harvard-Yenching Institute 2012.
11. J. M. Bloom. Paper before print. *The history and impact of paper in the Islamic world.* Yale University Press 2001.
12. B. Q. Lu, Y. J. Zhu, F. Chen. Highly flexible and nonflammable inorganic hydroxyapatite paper. *Chemistry – A European Journal* 2014, *20*, 1242–1246.
13. F. Chen, Y. J. Zhu. Multifunctional calcium phosphate nanostructured materials and biomedical applications. *Current Nanoscience* 2014, *10*, 465–485.
14. B. Q. Lu, Y. J. Zhu. One-dimensional hydroxyapatite materials: preparation and applications. *Canadian Journal of Chemistry* 2017, *95*, 1091–1102.
15. Y. J. Zhu. Nanostructured materials of calcium phosphates and calcium silicates: synthesis, properties and applications. *Chinese Journal of Chemistry* 2017, *35*, 769–790.
16. S. Pai, S. M. Kini, R. Selvaraj, A. Pugazhendhi. A review on the synthesis of hydroxyapatite, its composites and adsorptive removal of pollutants from wastewater. *Journal of Water Process Engineering* 2020, *38*, 101574.
17. M. Ibrahim, M. Labaki, J. M. Giraudon, J. F. Lamonier. Hydroxyapatite, a multifunctional material for air, water and soil pollution control: a review. *Journal of Hazardous Materials* 2020, *383*, 121139.
18. A. Fihri, C. Len, R. S. Varma, A. Solhy. Hydroxyapatite: a review of syntheses, structure and applications in heterogeneous catalysis. *Coordination Chemistry Reviews* 2017, *347*, 48–76.
19. A. Das, D. Pamu. A comprehensive review on electrical properties of hydroxyapatite based ceramic composites. *Materials Science and Engineering C* 2019, *101*, 539–563.
20. D. Arcos, M. Vallet-Regí. Substituted hydroxyapatite coatings of bone implants. *Journal of Materials Chemistry B* 2020, *8*, 1781–1800
21. P. Madhavasarma, P. Veeraragavan, S. Kumaravel, M. Sridevi. Studies on physiochemical modifications on biologically important hydroxyapatite materials and their characterization for medical applications. *Biophysical Chemistry* 2020, *267*, 106474.

22. F. Chen, Y. J. Zhu, K. W. Wang, K. L. Zhao. Surfactant-free solvothermal synthesis of hydroxylapatite nanowire/nanotube ordered arrays with biomimic structures. *CrystEngComm* 2011, *13*, 1858–1863.

23. X. Y. Zhao, Y. J. Zhu, F. Chen, B. Q. Lu, C. Qi, J. Zhao, J. Wu. Hydrothermal synthesis of hydroxyapatite nanorods and nanowires using riboflavin-5′-phosphate monosodium salt as a new phosphorus source and their application in protein adsorption. *CrystEngComm* 2013, *15*, 7926–7935.

24. S. Bramhe, T. N. Kim, A. Balakrishnan, M. C. Chu. Conversion from biowaste Venerupis clam shells to hydroxyapatite nanowires. *Materials Letters* 2014, *135*, 195–198.

25. M. Ai, Z. Du, S. Zhu, H. Geng, X. Zhang, Q. Cai, X. Yang. Composite resin reinforced with silver nanoparticles–laden hydroxyapatite nanowires for dental application. *Dental Materials* 2017, *33*, 12–22.

26. C. Qi, Q. L. Tang, Y. J. Zhu, X. Y. Zhao, F. Chen. Microwave-assisted hydrothermal rapid synthesis of hydroxyapatite nanowires using adenosine 5′-triphosphate disodium salt as phosphorus source. *Materials Letters* 2012, *85*, 71–73.

27. K. Lin, X. Liu, J. Chang, Y. J. Zhu. Facile synthesis of hydroxyapatite nanoparticles, nanowires and hollow nano-structured microspheres using similar structured hardprecursors. *Nanoscale* 2011, *3*, 3052–3055.

28. Z. Yang, Y. Huang, S. T. Chen, Y. Q. Zhao, H. L. Li, Z. A. Hu. Template synthesis of highly ordered hydroxyapatite nanowire arrays. *Journal of Materials Science* 2005, *40*, 1121–1125.

29. D. O. Costa, S. J. Dixon, A. S. Rizkalla. One- and three-dimensional growth of hydroxyapatite nanowires during sol–gel–hydrothermal synthesis. *ACS Applied Materials & Interfaces* 2012, *4*, 1490–1499.

30. M. Cao, Y. Wang, C. Guo, Y. Qi, C. Hu. Preparation of ultrahigh-aspect-ratio hydroxyapatite nanofibers in reverse micelles under hydrothermal conditions. *Langmuir* 2004, *20*, 4784–4786.

31. Y. Y. Jiang, Y. J. Zhu, F. Chen, J. Wu. Solvothermal synthesis of submillimeter ultralong hydroxyapatite nanowires using a calcium oleate precursor in a series of monohydroxy alcohols. *Ceramics International* 2015, *41*, 6098–6102.

32. Y. G. Zhang, Y. J. Zhu, F. Chen, J. Wu. Ultralong hydroxyapatite nanowires synthesized by solvothermal treatment using a series of phosphate sodium salts. *Materials Letters* 2015, *144*, 135–137.

33. H. P. Yu, Y. J. Zhu, B. Q. Lu. Highly efficient and environmentally friendly microwaveassisted hydrothermal rapid synthesis of ultralong hydroxyapatite nanowires. *Ceramics International* 2018, *44*, 12352–12356.

34. H. Li, Y. J. Zhu, Y. Y. Jiang, Y. D. Yu, F. Chen, L. Y. Dong, J. Wu. Hierarchical assembly of monodisperse hydroxyapatite nanowires and construction of high-strength fireresistant inorganic paper with high-temperature flexibility. *ChemNanoMat* 2017, *3*, 259–268.

35. F. Chen, Y. J. Zhu. Large-scale automated production of highly ordered ultralong hydroxyapatite nanowires and construction of various fire-resistant flexible ordered architectures. *ACS Nano* 2016, *10*, 11483–11495.

36. Y. J. Zhu, F. Chen. pH-responsive drug delivery systems. *Chemistry – An Asian Journal* 2015, *10*, 284–305.

37. R. L. Yang, Y. J. Zhu, F. F. Chen, D. D. Qin, Z. C. Xiong. Bioinspired macroscopic ribbon fibers with a nacre-mimetic architecture based on highly ordered alignment of ultralong hydroxyapatite nanowires. *ACS Nano* 2018, *12*, 12284–12295.

38. M. Y. Ma, Y. J. Zhu, L. Li, S. W. Cao. Nanostructured porous hollow ellipsoidal capsules of hydroxyapatite and calcium silicate: preparation and application in drug delivery. *Journal of Materials Chemistry* 2008, *18*, 2722–2727.

39. K. W. Wang, Y. J. Zhu, X. Y. Chen, W. Y. Zhai, Q. Wang, F. Chen, J. Chang, Y. R. Duan. Flower-like hierarchically nanostructured hydroxyapatite hollow spheres: facile preparation and application in anticancer drug cellular delivery. *Chemistry – An Asian Journal* 2010, *5*, 2477–2482.

40. F. Chen, P. Huang, Y. J. Zhu, J. Wu, C. L. Zhang, D. X. Cui. The photoluminescence, drug delivery and imaging properties of multifunctional Eu^{3+}/Gd^{3+} dual-doped hydroxyapatite nanorods. *Biomaterials* 2011, *32*, 9031–9039.

41. C. Qi, Y. J. Zhu, B. Q. Lu, X. Y. Zhao, J. Zhao, F. Chen. Hydroxyapatite nanosheet-assembled porous hollow microspheres: DNA-templated hydrothermal synthesis, drug delivery and protein adsorption. *Journal of Materials Chemistry* 2012, *22*, 22642–22650.

42. C. Qi, Y. J. Zhu, X. Y. Zhao, J. Zhao, F. Chen, G. F. Cheng, Y. J. Ruan. High surface area carbonate apatite nanorod bundles: surfactant-free sonochemical synthesis and drug loading and release properties. *Materials Research Bulletin* 2013, *48*, 1536–1540.

43. F. Chen, C. Li, Y. J. Zhu, X. Y. Zhao, B. Q. Lu, J. Wu. Magnetic nanocomposite of hydroxyapatite ultrathin nanosheets/Fe_3O_4 nanoparticles: microwave-assisted rapid synthesis and application in pH-responsive drug release. *Biomaterials Science* 2013, *1*, 1074–1081.

44. C. Qi, Y. J. Zhu, B. Q. Lu, X. Y. Zhao, J. Zhao, F. Chen, J. Wu. Hydroxyapatite hierarchically nanostructured porous hollow microspheres: rapid, sustainable microwave-hydrothermal synthesis by using creatine phosphate as an organic phosphorus source and application in drug delivery and protein adsorption. *Chemistry – A European Journal* 2013, *19*, 5332–5341.

45. W. L. Yu, T. W. Sun, Z. Y. Ding, C. Qi, H. K. Zhao, F. Chen, Z. M. Shi, Y. J. Zhu, D. Y. Chen, Y. H. He. Copper-doped mesoporous hydroxyapatite microspheres synthesized by a microwave-hydrothermal method using creatine phosphate as an organic phosphorus source: application in drug delivery and enhanced bone regeneration. *Journal of Materials Chemistry B* 2017, *5*, 1039–1052.

46. Y. D. Yu, Y. J. Zhu, C. Qi, J. Wu. Hydroxyapatite nanorod-assembled hierarchical microflowers: rapid synthesis via microwave hydrothermal transformation of $CaHPO_4$ and their application in protein/drug delivery. *Ceramics International* 2017, *43*, 6511–6518.

47. Y. D. Yu, Y. J. Zhu, C. Qi, Y. Y. Jiang, H. Li, J. Wu. Hydroxyapatite nanorod-assembled porous hollow polyhedra as drug/protein carriers. *Journal of Colloid and Interface Science* 2017, *496*, 416–424.

48. Y. G. Zhang, Y. J. Zhu, F. Chen, T. W. Sun, Y. Y. Jiang. Ultralong hydroxyapatite microtubes: solvothermal synthesis and application in drug loading and sustained drug release. *CrystEngComm* 2017, *19*, 1965–1973.

49. Y. G. Zhang, Y. J. Zhu, F. Chen, T. W. Sun. A novel composite scaffold comprising ultralong hydroxyapatite microtubes and chitosan: preparation and application in drug delivery. *Journal of Materials Chemistry B* 2017, *5*, 3898–3906.

50. A. Y. Cai, Y. J. Zhu, C. Qi. Biodegradable inorganic nanostructured biomaterials for drug delivery. *Advanced Materials Interfaces* 2020, *7*, 2000819.

51. C. Qi, S. Musetti, L. H. Fu, Y. J. Zhu, L. Huang. Biomolecule-assisted green synthesis of nanostructured calcium phosphates and their biomedical applications. *Chemical Society Reviews* 2019, *48*, 2698–2737.

52. D. Zhou, C. Qi, Y. X. Chen, Y. J. Zhu, T. W. Sun, F. Chen, C. Q. Zhang. Comparative study of porous hydroxyapatite/chitosan and whitlockite/chitosan scaffolds for bone regeneration in calvarial defects. *International Journal of Nanomedicine* 2017, *12*, 2673–2687.

53. W. L. Yu, T. W. Sun, C. Qi, Z. Y. Ding, H. K. Zhao, S. C. Zhao, Z. M. Shi, Y. J. Zhu, D. Y. Chen, Y. H. He. Evaluation of zinc-doped mesoporous hydroxyapatite microspheres for the construction of a novel biomimetic scaffold optimized for bone augmentation. *International Journal of Nanomedicine* 2017, *12*, 2293–2306.

54. W. L. Yu, T. W. Sun, C. Qi, H. K. Zhao, Z. Y. Ding, Z. W. Zhang, B. B. Sun, J. Shen, F. Chen, Y. J. Zhu, D. Y. Chen, Y. H. He. Enhanced osteogenesis and angiogenesis by mesoporous hydroxyapatite microspheres-derived simvastatin sustained release system for superior bone regeneration. *Scientific Reports* 2017, *7*, 44129.

55. B. Nemery. Occupational diseases-overview. *Encyclopedia of Respiratory Medicine* 2006, 186–191.

56. Z. C. Xiong, Z. Y. Yang, Y. J. Zhu, F. F. Chen, Y. G. Zhang, R. L. Yang. Ultralong hydroxyapatite nanowires-based paper co-loaded with silver nanoparticles and antibiotic for long-term antibacterial benefit. *ACS Applied Materials & Interfaces* 2017, *9*, 22212–22222.

57. F. F. Chen, Y. J. Zhu, Z. C. Xiong, T. W. Sun, Y. Q. Shen, R. L. Yang. Inorganic nanowires-assembled layered paper as the valve for controlling water transportation. *ACS Applied Materials & Interfaces* 2017, *9*, 11045–11053.

58. L. Y. Dong, Y. J. Zhu. A new kind of fireproof, flexible, inorganic, nanocomposite paper and its application to the protection layer in flame-retardant fiber-optic cables. *Chemistry – A European Journal* 2017, *23*, 4597–4604.

59. F. F. Chen, Y. J. Zhu, Z. C. Xiong, T. W. Sun, Y. Q. Shen. Highly flexible superhydrophobic and fire-resistant layered inorganic paper. *ACS Applied Materials & Interfaces* 2016, *8*, 34715–34724.

60. Z. C. Xiong, Y. J. Zhu, F. F. Chen, T. W. Sun, Y. Q. Shen. One-step synthesis of silver nanoparticle-decorated hydroxyapatite nanowires for the construction of highly flexible free-standing paper with high antibacterial activity. *Chemistry – A European Journal* 2016, *22*, 11224–11231.

61. F. F. Chen, Y. J. Zhu, Z. C. Xiong, L. Y. Dong, F. Chen, B. Q. Lu, R. L. Yang. Hydroxyapatite nanowire-based all-weather flexible electrically conductive paper with superhydrophobic and flame-retardant properties. *ACS Applied Materials & Interfaces* 2017, *9*, 39534–39548.

62. R. L. Yang, Y. J. Zhu, F. F. Chen, D. D. Qin, Z. C. Xiong. Recyclable, fire-resistant, superhydrophobic, and magnetic paper based on ultralong hydroxyapatite nanowires for continuous oil/water separation and oil collection. *ACS Sustainable Chemistry & Engineering* 2018, *6*, 10140–10150.

63. R. L. Yang, Y. J. Zhu, F. F. Chen, L. Y. Dong, Z. C. Xiong. Luminescent, fire-resistant, and water-proof ultralong hydroxyapatite nanowire-based paper for multimode anti-counterfeiting applications. *ACS Applied Materials & Interfaces* 2017, *9*, 25455–25464.

64. Z. C. Xiong, Z. Y. Yang, Y. J. Zhu, F. F. Chen, R. L. Yang, D. D. Qin. Ultralong hydroxyapatite nanowire-based layered catalytic paper for highly efficient continuous flow reactions. *Journal of Materials Chemistry A* 2018, *6*, 5762–5773.

65. Z. C. Xiong, Y. J. Zhu, D. D. Qin, F. F. Chen, R. L. Yang. Flexible fire-resistant photothermal paper comprising ultralong hydroxyapatite nanowires and carbon nanotubes for solar energy-driven water purification. *Small* 2018, *14*, 1803387.

66. R. L. Yang, Y. J. Zhu, F. F. Chen, D. D. Qin, Z. C. Xiong. Superhydrophobic photothermal paper based on ultralong hydroxyapatite nanowires for controllable light-driven self-propelled motion. *ACS Sustainable Chemistry & Engineering* 2019, *7*, 13226–13235.

67. R. L. Yang, Y. J. Zhu, D. D. Qin, Z. C. Xiong. Light-operated dual-mode propulsion at the liquid/air interface using flexible, superhydrophobic, and thermally stable photothermal paper. *ACS Applied Materials & Interfaces* 2020, *12*, 1339–1347.

68. S. X. Wu. *Shanxi JinZhiYuan Mural Art Museum. The history of Chinese Xuan paper.* Shanxi Economic Publishing 2016.

69. The Educational, Scientific and Cultural Organization of United Nations. Nomination File No. 00201. *The traditional handicrafts of making Xuan paper.* https://ich.unesco.org/en/RL/traditionalhandicrafts-of-making-xuan-paper-00201. 2009.

70. R. Q. Liu. The human heritage of the national treasure Xuan paper. *Encyclopedic Knowledge* 2010, *2*, 24–27.

71. W. J. Hu, D. S. Zhao. Analysis of the life characteristics of Xuan paper. *China Pulp & Paper Industry* 2013, *34*, 74–76.

72. L. Y. Dong, Y. J. Zhu. Fire-resistant inorganic analogous Xuan paper with thousands of years' super-durability. *ACS Sustainable Chemistry & Engineering* 2018, *6*, 17239–17251.

73. R. Q. Liu, Y. L. Qu. Permanency of Xuan paper-a preliminary study. *China Pulp and Paper* 1986, 32–37.

74. M. C. Area, H. Cheradame. Paper aging and degradation: recent findings and research methods. *BioResources* 2011, *6*, 5307–5337.

75. L. Y. Dong, Y. J. Zhu, Q. Q. Zhang, Y. T. Shao. Fire-retardant and high-temperature-resistant label paper and its potential applications. *ChemNanoMat* 2019, *5*, 1418–1427.

76. L. Y. Dong, Y. J. Zhu. Fire-retardant paper with ultrahigh smoothness and glossiness. *ACS Sustainable Chemistry & Engineering* 2020, *8*, 17500–17507.

77. F. F. Chen, Y. J. Zhu, F. Chen, L. Y. Dong, R. L. Yang, Z. C. Xiong. Fire alarm wallpaper based on fire-resistant hydroxyapatite nanowire inorganic paper and graphene oxide thermosensitive sensor. *ACS Nano* 2018, *12*, 3159–3171.

78. F. F. Chen, Y. J. Zhu, Q. Q. Zhang, R. L. Yang, D. D. Qin, Z. C. Xiong. Secret paper with vinegar as an invisible security ink and fire as a decryption key for information protection. *Chemistry – A European Journal* 2019, *25*, 10918–10925.

79. H. Eccles. Edited by: A. Dyer, M. J. Hudson and P. A. Williams. Ion exchange-future challenges/opportunities in environmental clean-up. *Progress in Ion Exchange-Advances and Applications* 1997, 245–259.

80. Q. Q. Zhang, Y. J. Zhu, J. Wu, Y. T. Shao, A. Y. Cai, L. Y. Dong. Ultralong hydroxyapatite nanowire-based filter paper for high-performance water purification. *ACS Applied Materials & Interfaces* 2019, *11*, 4288–4301.

81. Q. Q. Zhang, Y. J. Zhu, J. Wu, Y. T. Shao, L. Y. Dong. A new kind of filter paper comprising ultralong hydroxyapatite nanowires and double metal oxide nanosheets for high-performance dye separation. *Journal of Colloid and Interface Science* 2020, *575*, 78–87.

82. Q. Q. Zhang, Y. J. Zhu, J. Wu, L. Y. Dong. Nanofiltration filter paper based on ultralong hydroxyapatite nanowires and cellulose fibers/nanofibers. *ACS Sustainable Chemistry & Engineering* 2019, *7*, 17198–17209.

83. Z. C. Xiong, Y. J. Zhu, D. D. Qin, R. L. Yang. Flexible salt-rejecting photothermal paper based on reduced graphene oxide and hydroxyapatite nanowires for high-efficiency solar energy-driven vapor generation and stable desalination. *ACS Applied Materials & Interfaces* 2020, *12*, 32556–32565.

84. Z. C. Xiong, R. L. Yang, Y. J. Zhu, F. F. Chen, L. Y. Dong. Flexible hydroxyapatite ultralong nanowire-based paper for highly efficient and multifunctional air filtration. *Journal of Materials Chemistry A* 2017, *5*, 17482–17491.

85. H. Li, D. B. Wu, J. Wu, L. Y. Dong, Y. J. Zhu, X. L. Hu. Flexible, high-wettability and fire-resistant separators based on hydroxyapatite nanowires for advanced lithium-ion batteries. *Advanced Materials* 2017, *29*, 1703548.

86. H. Li, L. Peng, D. B. Wu, J. Wu, Y. J. Zhu, X. L. Hu. Ultrahigh-capacity and fire-resistant LiFePO$_4$-based composite cathodes for advanced lithium-ion batteries. *Advanced Energy Materials* 2019, *9*, 1802930.

87. H. Li, S. T. Guo, L. B. Wang, J. Wu, Y. J. Zhu, X. L. Hu. Thermally durable lithium-ion capacitors with high energy density from all hydroxyapatite nanowire-enabled fire-resistant electrodes and separators. *Advanced Energy Materials* 2019, *9*, 1902497.

88. Y.-J. Zhu, B.-Q. Lu. Deformable biomaterials based on ultralong hydroxyapatite nanowires. *ACS Biomaterials Science & Engineering* 2019, *5*, 4951–4961.

89. T. W. Sun, Y. J. Zhu, F. Chen. Highly flexible multifunctional biopaper comprising chitosan reinforced by ultralong hydroxyapatite nanowires. *Chemistry – A European Journal* 2017, *23*, 3850–3862.

90. P. Kithva, L. Grondahl, D. Martin, M. Trau. *Journal of Materials Chemistry* 2010, *20*, 381–389.

91. N. L. B. M. Yusof, A. Wee, L. Y. Lim, E. Khor. Flexible chitin films as potential wound-dressing materials: wound model studies. *Journal of Biomedical Materials Research A* 2003, *66A*, 224–232.

92. T. W. Sun, Y. J. Zhu, F. Chen, Y. G. Zhang. Ultralong hydroxyapatite nanowire/collagen biopaper with high flexibility, improved mechanical properties and excellent cellular attachment. *Chemistry – An Asian Journal* 2017, *12*, 655–664.

93. H. Kashiwada, Y. Shimizu, Y. Sano, K. Yamauchi, H. Guang, H. Kumamoto, H. Unuma, Y. J. Zhu. In vivo behaviors of highly flexible paper consisting of ultralong hydroxyapatite nanowires. *Journal of Biomedical Materials Research B* 2021, *109*, DOI: 10.1002/jbm.b.34819.

94. Z. F. Zhou, T. W. Sun, Y. H. Qin, Y. J. Zhu, Y. Y. Jiang, Y. Zhang, J. J. Liu, J. Wu, S. S. He, F. Chen. Selenium-doped hydroxyapatite biopapers with an anti-bone tumor effect by inducing apoptosis. *Biomaterials Science* 2019, *7*, 5044–5053.

95. F. F. Chen, Y. J. Zhu, Z. C. Xiong, T. W. Sun. Hydroxyapatite nanowires@metal-organic framework core/shell nanofibers: templated synthesis, peroxidase-like activity, and derived flexible recyclable test paper. *Chemistry – A European Journal* 2017, *23*, 3328–3337.

96. J. Rojas-Torres, M. Cea, Y. J. Zhu, G. M. Fonseca. Behavior of 4 types of paper with printed QR codes for evaluating denture marking in conditions of extreme heat. *The Journal of Prosthetic Dentistry* 2021, *124–125*, DOI: 10.1016/j.prosdent.2020.08.032.

97. F. F. Chen, Y. J. Zhu, Y. G. Zhang, R. L. Yang, H. P. Yu, D. D. Qin, Z. C. Xiong. Portable and writable photoluminescent chalk for on-site information protection on arbitrary substrates. *Chemical Engineering Journal* 2019, *369*, 766–774.

98. F. F. Chen, Z. Y. Yang, Y. J. Zhu, Z. C. Xiong, L. Y. Dong, B. Q. Lu, J. Wu, R. L. Yang. Low-cost and scaled-up production of fluorine-free, substrate-independent, large-area superhydrophobic coatings based on hydroxyapatite nanowire bundles. *Chemistry – A European Journal* 2018, *24*, 416–424.

99. T. W. Sun, Y. J. Zhu. Solvothermal growth of ultralong hydroxyapatite nanowire coating on glass substrate. *Chemistry Letters* 2019, *48*, 1462–1464.

100. H. P. Yu, Y. J. Zhu, B. Q. Lu. Dental enamel-mimetic large-sized multi-scale ordered architecture built by a well controlled bottom-up strategy. *Chemical Engineering Journal* 2019, *360*, 1633–1645.

101. Y. G. Zhang, Y. J. Zhu, Z. C. Xiong, J. Wu, F. Chen. Bioinspired ultralight inorganic aerogel for highly efficient air filtration and oil-water separation. *ACS Applied Materials & Interfaces* 2018, *10*, 13019–13027.

102. G. J. Huang, H. P. Yu, X. L. Wang, B. B. Ning, J. Gao, Y. Q. Shi, Y. J. Zhu, J. L. Duan. Highly porous and elastic aerogel based on ultralong hydroxyapatite nanowires for high-performance bone regeneration and neovascularization. *Journal of Materials Chemistry B* 2021, *9*, 1277–1287.

103. T. W. Sun, Y. J. Zhu, F. Chen. Hydroxyapatite nanowire/collagen elastic porous nano-composite and its enhanced performance in bone defect repair. *RSC Advances* 2018, *8*, 26133–26144.

104. T. W. Sun, Y. J. Zhu, F. Chen, F. F. Chen, Y. Y. Jiang, Y. G. Zhang, J. Wu. Ultralong hydroxyapatite nanowires/collagen scaffolds with hierarchical porous structure, enhanced mechanical properties and excellent cellular attachment. *Ceramics International* 2017, *43*, 15747–15754.

105. T. W. Sun, W. L. Yu, Y. J. Zhu, F. Chen, Y. G. Zhang, Y. Y. Jiang, Y. H. He. Porous nanocomposite comprising ultralong hydroxyapatite nanowires decorated with zinc-containing nanoparticles and chitosan: synthesis and application in bone defect repair. *Chemistry – A European Journal* 2018, *24*, 8809–8821.

106. T. W. Sun, Y. J. Zhu. One-step solvothermal synthesis of strontium-doped ultralong hydroxyapatite nanowires. *Journal of Inorganic Materials* 2020, *35*, 724–728.

107. T. W. Sun, W. L. Yu, Y. J. Zhu, R. L. Yang, Y. Q. Shen, D. Y. Chen, Y. H. He, F. Chen. Hydroxyapatite nanowire@magnesium silicate core-shell hierarchical nanocomposite: synthesis and application in bone regeneration. *ACS Applied Materials & Interfaces* 2017, *9*, 16435–16447.

108. Y. G. Zhang, Y. J. Zhu, F. Chen, T. W. Sun, Y. Y. Jiang. Highly porous ceramics based on ultralong hydroxyapatite nanowires. *RSC Advances* 2016, *6*, 102003-102009.

109. Y. Y. Jiang, Y. J. Zhu, H. Li, Y. G. Zhang, Y. Q. Shen, T. W. Sun, F. Chen. Preparation and enhanced mechanical properties of hybrid hydrogels comprising ultralong hydroxy-apatite nanowires and sodium alginate. *Journal of Colloid and Interface Science* 2017, *497*, 266–275.

110. J. Wu, Y. J. Zhu. Acid/alkali-proof fire-resistant inorganic paper comprising barium sulfate nanorods-assembled fibers. *European Journal of Inorganic Chemistry* 2021, *2021*, 492–499.

Index

Taylor & Francis eBooks

www.taylorfrancis.com

A single destination for eBooks from Taylor & Francis
with increased functionality and an improved user
experience to meet the needs of our customers.

90,000+ eBooks of award-winning academic content in
Humanities, Social Science, Science, Technology, Engineering,
and Medical written by a global network of editors and authors.

TAYLOR & FRANCIS EBOOKS OFFERS:

A streamlined
experience for
our library
customers

A single point
of discovery
for all of our
eBook content

Improved
search and
discovery of
content at both
book and
chapter level

REQUEST A FREE TRIAL
support@taylorfrancis.com

 Routledge
Taylor & Francis Group

 CRC Press
Taylor & Francis Group

Printed in the United States
by Baker & Taylor Publisher Services

Printed in the United States
by Baker & Taylor Publisher Services